FEVERFEW

DR STEWART JOHNSON is a pharmacologist and medical practitioner. He is an Honorary Consultant at The City of London Migraine Clinic. He belongs to the British Pharmacological Society, the Physiological Society, the British Medical Association and the Brain Research Association, and is a former Director of the King's College Centre of Law, Medicine and Ethics and Reader in Pharmacology at King's College. He has written numerous papers and articles on various aspects of medicine, and has talked about feverfew on both television and radio. He is married, with two children.

Overcoming Common Problems Series

Overcoming Common Problems Series

Overcoming Common Problems

FEVERFEW

Dr Stewart Johnson
PhD, BSc, MB, BS

SHELDON PRESS
LONDON

First published in Great Britain in 1984 by
Sheldon Press, SPCK, Marylebone Road, London NW1 4DU
Second impression 1986

Thanks are due to Gerald Duckworth and Co. Ltd. for permission
to quote four lines from *Cautionary Tales*, *Henry King*
by Hilaire Belloc.

British Library Cataloguing in Publication Data

Johnson, Stewart
    Feverfew.—(Overcoming common problems)
    1. Migraine   2. Feverfew—Therapeutic use
    I. Title   II. Series
    616.8'57061   RC392

    ISBN 0–85969–423–2
    ISBN 0–85969–424–0 Pbk

Typeset by Inforum Ltd, Portsmouth
Printed in Great Britain by
Richard Clay (The Chaucer Press) Ltd
Bungay, Suffolk

Physicians of the Utmost Fame
Were called at once; but when they came
They answered, as they took their Fees,
There is no Cure for this Disease.

HILAIRE BELLOC, 1870—1953
Cautionary Tales, Henry King

Contents

Acknowledgements

The letters used in this book to demonstrate the effectiveness or side-effects of feverfew represent just a tiny proportion of the unsolicited communications received from feverfew-users. I am most grateful to the many correspondents who have written to me over the past five years.

I am also indebted to all those feverfew users whose co-operation made the series of clinical investigations possible, and in particular to Mrs Ann Jenkins and Dr J R Gledhill for providing names and addresses of feverfew-users and enquirers, and to Mrs Lesley Hirst for her pill 'recipe' and instructions to other users. Without their help and the secretarial assistance of Mrs Mary Farey, the clinical assistance of Mrs Nirmala Kadam and Sister Joan Vincent at the City of London Migraine Clinic and the technical support of Miss Marilyn Biggs in encoding the results, this work would have taken much longer.

I also owe debts of gratitude to my colleague Miss Margaret Skinner of the King's College Computer Centre who computerized the feverfew results and was the source of endless statistical advice, and to Drs Deborah Jessup and Peter Hylands of Chelsea College who provided tablets and capsules for the clinical studies mentioned and prepared the extracts of feverfew for biological testing.

My colleagues, the joint Medical Directors of the City of London Migraine Clinic, Drs Marcia Wilkinson and Nat Blau have also generously given both advice and encouragement throughout the feverfew investigations.

The cost of the data transcription from questionnaires to Fortran code and some of the expenses incurred by trial patients in coming to London were borne by the British Migraine Association, whose support is greatly appreciated.

Finally, but most importantly, I wish to thank my wife Dr Trilby Johnson for her help in drafting the questionnaires and for encouraging me to write this account. She has now begun her own feverfew study.

Introduction

The Thalidomide tragedy of the early sixties highlighted deficiencies in the ways new medical substances were assessed for safety prior to marketing. The public was understandably outraged and, as a result, the 1968 *Medicines' Act* was introduced in Great Britain, to control the efficacy, safety and quality of all new drugs. In effect, the act has ensured that only those drugs that are safe and effective in common conditions ever appear.

It now costs over £40 million and more than ten years of experimentation for a major international drug company to develop a new medicine. And to ensure profitability the manufacturer must own the patent rights so that there will be no competition. It therefore requires an exceptional interest by a pharmaceutical company of repute in medicinal plants because they are extremely difficult to protect by patent.

However, we now know that, regardless of the rigorous scientific safeguards and of the stature of the independent assessors at the Department of Health and Social Security (DHSS), hazardous drugs continue to slip through the net. Oral contraceptives have caused dangerous blood clots, the heart drug Eraldin impaired eyesight. More recently, patients have died while taking the antiarthritis drug Opren, and, like the antidepressant drug Zelmid, the pain-killers Zomax, Osmosin and Flosint have been withdrawn.

Consequently, more and more patients are turning to 'natural' medicines, plant products that were the principal medicines used by our great-grandparents. 'Green pharmacy' is a stock phrase now used by health-care professionals to denote these herbal products, which are to be found in most chemists.

The sophisticated way in which our forebears used plants for their curative powers indicated a remarkable facility for careful observation, yet many were used for conditions that in part escaped the attention of the writers of the great herbals. Instead, these remedies were handed on from doctor to doctor, doctor to patient, patient to patient – true folk medicine.

FEVERFEW

Such is the story of the herb feverfew – the rediscovery by patients of a very old remedy for two extremely common conditions. Many hundreds of sufferers are convinced that their relative freedom from the painful symptoms of migraine and arthritis is mainly due to this plant. Yet, largely because of the patent difficulties mentioned above, feverfew has never been tested in animals in the way that would satisfy a regulatory body such as the UK Committee on Safety of Medicines.

1
Migraine Observed

Most of us have the occasional headache and usually it is of so little consequence we don't even bother to see our doctor. Headaches seem inextricably linked with life itself. Yet nobody seems to know *why* man suffers from them. In a few cases they are the warning signs of more serious abnormalities within the skull such as a growth, blood loss or infection, but probably the vast majority of severe headaches are due either to tension or migraine. Tension headache is so named because of the increased tension that occurs within the muscles of the scalp, neck and face; it often recurs day after day for weeks on end. Migraine is quite different.

What is migraine?

Migraine tends to run in families and is characteristically a severe headache – often on one side of the head only – which comes and goes, with days, weeks or even months of complete freedom between attacks. But it is not simply a severe headache. To be migraine, it is generally accepted by doctors that this intermittent headache has to be associated with other symptoms such as nausea, vomiting or diarrhoea. Before the headache, there may also be disturbances of the nervous system – such as visual symptoms, or numbness or weakness of an arm or leg on the opposite side of the body from the headache.

There are several kinds of migraine and the same person may experience different symptoms from attack to attack. Furthermore, no two people have attacks which are identical in form, frequency of occurrence, duration, or severity, and it is often only when the attack rate or severity changes that the sufferer is prompted to seek medical advice.

One of the interesting things about migraine is that nothing organically wrong is found on examination. As my former teacher and now colleague and perhaps this country's most knowledgeable migraine specialist, Dr Marcia Wilkinson, always impressed upon me: 'If you find anything wrong, it isn't migraine!'

There are three main types of migraine: classical migraine, common migraine and migrainous neuralgia or cluster headache.

Classical migraine

In classical migraine, before the headache the sufferer experiences symptoms such as flashing lights, geometric designs resembling the top of battlements, blurred sight, temporary partial loss of vision or even loss of speech. This stage, known as the aura, usually lasts only fifteen to twenty minutes. The patient usually dislikes being in daylight, preferring a quiet, darkened room. As the visual symptoms lessen, the headache begins, and gradually increases in intensity. It has a throbbing character and at its peak is a very severe nagging headache.

As the headache builds up the patient feels sick and very often will vomit, after which the pain sometimes eases.

Common migraine

This most frequently encountered form of migraine does not involve visual disturbances. The headache is often the first symptom to be perceived but it is just as severe as the classical migraine headache. It is as likely to be one-sided and usually returns on the same side of the head, though it *may* switch to the opposite side for no obvious reason. Nausea nearly always occurs and vomiting sometimes comes in addition, or shortly after. Both classical and common migraine are more often suffered by women than men; only a quarter to one third of all migraine sufferers are men.

Migrainous neuralgia (cluster headache)

This much less common form of migraine differs from classical and common migraine in that it usually affects men and only rarely women. The reason for this is not known. It is usually one-sided but the attacks tend to occur in groups or clusters which can last weeks or months and then suddenly disappear for several months or even years. The intensely painful headache comes on suddenly, lasts for one to one-and-a-half hours and may be at its most severe in, around or behind one eye, which waters and reddens. The nose feels 'stuffed up' on the affected side. The patient may get more than one attack a day and they usually occur on most days until, just as suddenly as they began, they cease.

When does it begin?

When 300 migraine sufferers, who had taken or were taking the herb feverfew were asked how old they were when their migraines began, more than half said that they began before the age of twenty and more than a quarter said they began at or before the age of twelve. Only one in ten said that attacks started after the age of forty-five.

This demonstrates one of the interesting characteristics of migraine: it usually begins in youth. Any incidence of severe headache coming on for the first time in late middle-age is taken very seriously and usually investigated in a hospital out-patient department.

So, the migraine patient is essentially one who has had recurrent attacks for a long time, during which there will have been opportunities for trying many of the conventional treatments available.

Timing, duration and frequency

Doctors (quite rightly) tell their patients to take pain-killers as soon as they detect the first symptoms of an impending attack. Unfortunately, this is often not possible because most sufferers from common migraine wake up with the headache already established and pain-killers don't seem to work very well when the migraine has already reached its peak. Migraine attacks can, however, occur at any time of day and some patients find that the symptoms come on in the evening just when they are beginning to relax after a hard day's work. Many patients report that attacks occur at weekends, when they tend to interfere with the whole family's leisure pursuits.

Some women suffer attacks regularly just before or with their monthly periods. And, somewhat surprisingly, if they are taking the contraceptive pill containing the two artificial hormones oestrogen and progestogen, the migraine attacks occur more often in the pill-free seven days. The commonly experienced remission during pregnancy and the menopause completes the trio of observations linking migraine with hormonal state in women.

In classical migraine the aura phase usually lasts only about a quarter of an hour but the headache which follows may last anything from sixty minutes to sixty hours. However, if anti-nauseants, pain-killers, or specific antimigraine drugs are taken very early in an attack (preferably with the first warning signs), or if the patient can get to sleep, the severity and duration are often curtailed.

Doctors feel uneasy about diagnosing migraine if the attacks occur more frequently than twice or three times a week as, by definition, migraine is an intermittent disorder with periods of complete normality in between. Except in migrainous neuralgia, migraine does *not* occur every day. Daily headaches are nearly always due to tension.

What is the cause?

No-one has yet adequately explained the cause, or causes of migraine. The throbbing, deep-seated pain, which is made worse by sudden movements such as standing upright after tying a shoelace, indicates that blood vessels inside the skull are involved. It is, in fact, known that blood vessels become narrower or wider during a migraine attack – but it is not easy to link these changes with the mainly nervous symptoms that occur.

The aura of classical migraine usually involves visual symptoms consisting of flashing lights, shimmering effects or partial loss of vision. These symptoms arise in the back of the brain (the visual cortex). When blood flow through this area has been measured while the patient is suffering the aura, a decrease has been observed, indicating a narrowing of the blood vessels. If very marked, this can so reduce the blood supply that the brain doesn't receive enough oxygen to allow the brain cells to function normally. Some doctors believe that this accounts for the visual disturbances.

The back of the brain is not the only part to be affected. The side of the brain that controls speech, sensation and movement can also suffer a temporary reduction in its blood supply. When this is marked, speech may become slurred, one hand and arm may feel numb or tingle with pins and needles and, in a few instances, the patient may lose the ability to grip properly or even notice that she is dragging one foot. All these symptoms last for only a few minutes – which is surprising because research in Copenhagen has disclosed that the narrowing of the blood vessels persists for several hours after the aura has ended. For some reason, still not fully understood, the blood vessels become very sensitive and when this happens the headache begins.

One of the puzzles of migraine is that while the narrowing of the blood vessels may occur simultaneously in both sides of the head, the migrainous headache is one-sided in about seven out of ten patients.

For several decades, research workers have been interested in what causes these blood vessel changes. They have investigated whether any chemical substances in the blood alter during the attack, which might account for the changes in blood flow to the brain, and have suggested that increases in certain natural substances called 'amines' and 'prostaglandins' might be responsible. This is because some blood vessels are known to narrow or widen in the presence of these substances.

The chief amines thought to be responsible are serotonin, noradrenaline and histamine: blood levels of serotonin and noradrenaline rise early in a migraine attack and blood levels of histamine increase in patients with migrainous neuralgia.

As the level of these substances may increase in stressful situations it is not surprising that migraine patients report that attacks often occur when they are under mental or physical stress. To oppose the actions of these substances in the body anti-drugs have been developed: these are the antiserotonins, antihistamines and antinoradrenalines.

Prostaglandins are believed to be involved in the production of pain, whether the head pain of migraine or the joint pain of arthritis. Although no true antiprostaglandin drug has yet been introduced for general use, drugs now known to be capable of preventing the formation of prostaglandins – aspirin for instance – have been in use for almost a century. The discovery of how aspirin works was made largely by British researchers, one of whom was awarded the Nobel Prize for Medicine in 1982.

Does anything provoke an attack of migraine?

When 300 migraine sufferers who were taking, or had taken, feverfew, were asked whether anything in particular brought on a migraine attack, eighty-three per cent mentioned one or several of the following, which are given in order of frequency of reporting:

	%		%
Stress (anxiety and tension)	37.4	Sudden loud noise; fright	7.2
Chocolate or cocoa	32.3		
Cheese and dairy products	28.1	Strong odours; pleasant or otherwise	5.5
Alcohol	20.9		
Menstrual periods (% women)	18.9	Low blood sugar	5.5
Overwork or exercise	17.4	Relaxation	5.5
Citrus fruit	13.5	Tea or coffee	5.1
Extreme climatic conditions (intense heat; heavy rain; snow)	11.5	Onions	5.1
		Yeast extracts	3.8
Bright or flickering light; TV	11.5	Late rising	3.8
Smokey/stuffy atmosphere	8.5	Overeating	2.1
Long journeys	8.0	Fish	2.1
Fried foods	7.7	Sudden change in posture	1.7
Food preservatives (sodium nitrate); preserved foods	7.7	Draughts	1.0
		Allergy	0.5

All of these are well-known to doctors as 'trigger' factors and the proportion of 'feverfew-eating' migraine sufferers with trigger factors is very similar to that of 'non-feverfew-taking' migraine patients.

The trigger factors can be subdivided into four groups:
1 *Stress-related* factors consisting of (a) psychological (anger, anxiety, worry, elation, depression) and (b) physical (overwork, exercise, long journeys) stress

8

2 *Dietary triggers* comprising (a) foods such as chocolate, cocoa, oranges, fries, onions, fish, yeast extracts, tea, coffee (b) alcoholic drinks, especially wines (c) going without meals (low blood sugar)

3 *Hormonal factors* such as menstruation and the contraceptive pill

4 *Environmental factors* such as extremes of climatic conditions, bright light, flicker (fluorescent lights and TV, cinema), loud noise, smokey or stuffy atmosphere, strong odours etc.

As a result of the pioneering work of the British doctor Edda Hanington we now have a very plausible explanation as to why certain foodstuffs trigger migraine attacks. They contain large amounts of amines which are chemically similar to those produced in the body during stress and during a migraine attack (see page 7). The migraine sufferer seems to be particularly sensitive to these dietary amines.

For a detailed discussion of trigger factors, the reader is referred to the excellent account given in Dr Marcia Wilkinson's book *Living with Migraine*, (Heinemann Health Books, London).

Treatment of migraine

Avoidance of triggers

If there is a recognizable trigger factor or activity that provokes an attack, this should be avoided whenever possible. Of course, it isn't always practicable to avoid every trigger, but they can be kept to a minimum. Actually, some migraine sufferers go to the extreme of avoiding all foods believed to trigger migraine. Such faddiness only makes life difficult. In my view, patients should never avoid a particular foodstuff permanently unless they are absolutely certain that it provokes their attacks.

Treatment of the acute attacks

The patient who experiences only two or three migraine attacks a year often requires only simple pain-killers such as aspirin and paracetamol. If nausea and vomiting are a problem, then an antisickness medication such as metoclopramide (Maxolon or Primperan) or prochlorperazine (Stemetil) is also prescribed, and should be taken as soon as the first symptom of an impending attack is noted. In cases where pills make the patient more likely to vomit, the medication can be given as suppositories.

If the migraine does not respond to simple pain-killers, more powerful drugs such as ergotamine may be prescribed. The disadvantage of ergotamine is that it can produce serious side-effects if the stated dose is exceeded.

If psychological stress is an obvious trigger, sedatives may be prescribed. The preferred sedatives are those which enable the patient to get off to sleep without subsequently causing persistent muzziness. Temazepam (Normison) is a useful tranquillizer in this situation, as it remains in the body for only a short time.

Prevention with drugs

When patients suffer more than two attacks every month, drugs which are antiserotonin in action, such as pizotifen, or antinoradrenaline in action, such as propranolol, may be prescribed to prevent attacks. The trouble is, although they are of benefit to many, it is very difficult to get patients to take them every day. Furthermore, a significant minority experiences side-effects when taking these drugs.

Whenever migraine is tackled with a new preventive treatment a proportion of sufferers experiences a marked improvement for two to three months, then the attacks return to their previous frequency, severity and duration. The patient, whose hopes had been raised initially, will then ask to be taken off the tablets. This initial favourable response to treatment is very common, perhaps as high as thirty-five per cent, especially if the patient is taking part in a clinical trial of a new medicine. This is known as a placebo-reaction (from the Latin placebo, meaning I shall please). For this reason it is essential to compare a new drug with a dummy (placebo) to make sure that the initial placebo benefit is not wrongly credited to the new drug.

Scope for improvement

Migraine is one of the commonest conditions afflicting mankind, in frequency of occurrence probably equivalent to hypertension or high blood pressure. So it is of interest to compare the amount spent by the UK National Health Service on drugs used to treat the two conditions. The annual bill for migraine treatments to the end of July 1983 came to £9.3 million whereas that for hypertension was £205 million. Why this enormous discrepancy? There is some overlap in that certain drugs used in migraine prevention – propranolol for example – are also used for hypertension, but this can account for only a very small proportion of the difference. Migraine patients also spend a lot of money on 'over the counter' medicines.

Probably the main reason why antimigraine drugs are not 'selling' is because, for the amount of relief gained, the incidence of side-effects is too high to make it worth while for patients to take these preventive medications once, twice or three times every day. After all, a patient whose frequency of severe migraine attacks is cut down from four attacks to two or three attacks a month is still getting severe migraine. Migraine is therefore one of the common medical conditions offering pharmaceutical companies considerable scope for improving on existing products.

Unfortunately, a pharmacological breakthrough in the treatment of migraine is not just around the corner; the underlying causes are so incompletely understood that it is more likely that any major new discovery will be made quite by chance.

Side-effects of antimigraine drugs

Some improvements have been made in the newer antimigraine treatments – side-effects, for instance, have been lessened – but there is still much scope for improvement. So what *are* the troublesome side-effects of commonly-used drugs?

Aspirin

Aspirin is among the most useful pain-killers and very difficult to improve on in migraine. Unfortunately it can also irritate the stomach lining, causing indigestion, stomach pain and, occasionally, bleeding of the stomach wall. It should not be used, therefore, by anyone suspected of having, or with a history of stomach and duodenal ulcers. Stomach upset tends to be lessened if aspirin is taken with food but eating is the last thing a patient suffering from acute migraine feels like doing. Effervescent aspirin, dissolved in water, is possibly easier on the stomach lining, but the volume that has to be taken tends to make some patients queasy.

Paracetamol

Paracetamol is a useful alternative if the patient cannot take aspirin. It does not irritate the stomach very often, although some patients find the large size of the tablets makes them difficult to swallow. Also, large doses over a long time may cause kidney damage and deliberate overdose may cause the liver to cease functioning altogether.

Caffeine

This is the nervous system stimulant present in tea and coffee. It is occasionally included in antimigraine treatments to help drugs like ergotamine to be absorbed more effectively, but I suggest patients avoid it whenever possible as it tends to keep them awake when they most need to get off to sleep. Caffeine-containing drinks can actually *provoke* migraine in some individuals.

Ergotamine (Femergin, Lingraine, Cafergot)

This drug is a 'natural' product derived from a fungus called ergot that grows on cereals. It often causes nausea and vomiting whichever way it is taken, and is sometimes combined with an antinauseant. A few patients get diarrhoea from the suppository formulation of ergotamine. Its prolonged use causes the hands and feet to become very cold due to the fact that blood vessels throughout the extremities become narrow. Ergotamine should never be taken for prolonged periods as a preventive because of this potentially dangerous side-effect. It also occasionally causes muscular pain and weakness in the legs.

Some doctors, having noted the extremely variable response to ergotamine due to poor absorption in a significant proportion of patients, believe that ergotamine should be scrapped. My own view is that it has a useful, if limited, role in the treatment of a minority of severely afflicted migraine sufferers. It should not be used for other causes of headache.

Antisickness drugs

Nausea and vomiting are frequently such troublesome symptoms of migraine that the patient finds them less bearable than the headache. A number of antisickness drugs are now available, one of the most commonly used being metoclopramide (Maxolon, Primperan) which, as well as stopping the nausea, causes the stomach to propel its contents along more efficiently, enabling pain-killing drugs to be absorbed more effectively.

Metoclopramide is usually free from untoward symptoms but in a small minority of adults, and more commonly in children – for whom it is therefore rarely prescribed – it causes distressing involuntary movements of the head and neck with grotesque facial grimacing and rhythmic protrusion of the tongue. Domperidone (Motilium), a newer antisickness drug, is said to have less of the side-effects of metoclopramide. Its place in migraine treatment is still being evaluated.

Tranquillizers

Drugs of the diazepam (Valium) kind may help patients through a short period of stress, but cease to be helpful in migraine if used for prolonged periods. They are habit forming and may cause day-time sedation so users should be warned not to drive cars or operate machinery. If patients can manage without them it is advisable that they should. However, the odd dose taken to induce sleep during an acute migraine attack can be beneficial.

Dihydroergotamine

This is a derivative of ergotamine but affects blood vessels less and can be used as a preventive. In high dosage its side-effects are similar to those of ergotamine. It opposes some of the actions of serotonin and noradrenaline.

Clonidine (Dixarit)

This drug apparently works in a few patients but may give rise to troublesome day-time drowsiness even in the low doses used to prevent migraine. It should not be stopped abruptly as sometimes in this situation the patient becomes agitated, cannot sleep, is nauseated and blood pressure may rise to dangerous levels. No one knows how clonidine actually works.

Propranolol (Inderal)

This opposes some of the actions of the amine noradrenaline (see page 7) released during the pre-headache phase of migraine. In the usual dosage it is relatively free from side-effects, although it can cause narrowing of the air passages in the lung in patients who suffer from asthma. It also slows the pulse and can precipitate heart failure in the elderly with heart disease. It often causes cold hands and feet and may give rise to vivid dreams and hallucinations. Even so, propranolol is possibly one of the most useful of the migraine preventive drugs.

Methysergide (Deseril)

This is another ergotamine derivative, introduced for its ability to oppose the actions of the amine, serotonin. As well as causing nausea, vomiting, heartburn and stomach pain, it can cause fluid retention in the body and, if used for more than six months, fibrous tissue may start to grow around the hollow organs of the body, restricting their function. The occurrence of this side-effect requires its immediate withdrawal. It is therefore better for methysergide to be prescribed only when all other treatments have been unsuccessful.

Pizotifen (Sanomigran)

This newest drug with antihistamine and antiserotonin actions is becoming increasingly popular and seems to benefit a good proportion of those for whom is is prescribed. It may cause day-time drowsiness, but this is apparently lessened if the daily dose is taken at bedtime. Patients should be warned by their doctors to expect an increase in appetite when on pizotifen – and consequent weight gain if food intake is not watched carefully. A few patients may also feel dizzy or nauseated.

Self help and the last hope

There are a great many ways in which the migraine sufferer can help herself. Learning to relax, avoiding trigger factors (see page 8), finding the pain-killer that suits her best, and lying down in a quiet room are some of them. But judging from the correspondence I have received, going to see the doctor is often not high on the list. This may be because, as patients seldom feel well enough to see their doctors while they are having a migraine attack, when they are better there seems no reason to trouble him. The last thing a patient wants to do during an attack, of course, is to make an appointment and have to drive or walk to the surgery. Being aware that her condition is only temporary and that she will be back to normal in a day or so, makes her disinclined to request a home visit by the family doctor.

A recent survey by a major drug company showed that some four out of every five migraine patients buy expensive 'over the counter' proprietary medicines to help them cope with their condition. Bearing in mind that an estimated one-eighth of the adult population suffers from migraine, this is self-medication on a very large scale indeed.

Does it perhaps indicate a lack of confidence on the part of some patients in the more migraine-specific drug treatments such as those containing ergotamine, previously prescribed by their doctor? This may be so, but doctors know only too well that many migraine patients who claim that a treatment does not work, have not taken the medicine in either the correct dose or, more importantly, at the correct time – that is, with the onset of the first symptoms of an impending attack. However, there are undoubtedly patients who, not having had any success with treatments obtained from their doctors, will try anything and everything the chemist has to offer. This can be very expensive and perhaps futile.

There are migraine sufferers who have tried nearly all of the available specific antimigraine drugs and preventives prescribed, taking them as directed and complying with the treatment in every way, often for many years – they have visited specialist migraine clinics and yet have gained no significant relief, or the drugs prescribed have caused unacceptable side-effects. Among these are women whose attacks seem to be linked with their menstrual cycle, occurring just before or on the first day of their periods. Some find that the only time they do not suffer these migraines is when they are pregnant. But pregnancy cannot be advocated as a treatment!

These patients can be so seriously disabled that they will try anything – such are the patients who eventually try feverfew, their last hope.

A typical story is given by a housewife in Tavistock:

> I have been a severe migraine sufferer since the age of eight years and have tried various treatments which made me vomit so much that I developed hiatus hernia. I am now unable to use any form of treatment containing the drug ergotamine as I developed side-effects, e.g. loss of feeling in toes and fingers and severe pain in the limbs. As you can imagine, I was more than pleased to read any information about cures for migraine. A friend found feverfew growing in a hedge. I started taking the leaf every day . . . I have since had no attacks until yesterday but . . . no vomiting took place . . . To be free from an attack of migraine is marvellous as I had one at least every week.

Another, who obtained only partial relief from aspirin or paracetamol was a Chislehurst woman, who wrote in October 1981:

> Since adolescence I had suffered migraine headaches on a more or less permanent basis, each headache lasting three to four days with only a day or two break before the next one. The only relief I got was from taking aspirin or paracetamol, the two having to be alternated at varying intervals as I seemed to get used to them after a while. I think on average I used to get through forty to fifty tablets per week!
>
> When I read about feverfew I really thought no more than 'oh well, I'll try anything once'. I knew when I started the leaves that the effects were not instant and would take anything up to six months to become fully effective, so perseverance was the key word, especially as I found the leaves disgusting to eat. Anyway, after two or three weeks I did notice my headaches were possibly lessening, but I used to

get fleeting 'twinges' to remind me of what it used to be like. After about three months my misery had almost completely disappeared, and now I just keep taking my daily feverfew to live a normal life.

Again, a sixty-year-old woman from Berkshire who had suffered two to three migraine attacks a week since childhood began taking two leaves of feverfew daily in June 1971. Five months later she wrote:

Since then I have experienced only one very minor migraine attack. All through my school life and right until the beginning of July this year, at least one day a week was sheer misery.

I have received literally hundreds of similar histories, often incorporating their past medical treatment which had not helped, as from a Bedworth woman:

I have had Dixarit, Sanomigran, Equagesic, Migraleve etc. which do not help me. I tried eating the leaves of the feverfew plant which I have in my garden every day, dipped in sugar . . . I don't know if it is coincidence but I have not had migraine for the last three months . . .

It was press reports of case histories such as these and the gradual realization that feverfew was apparently being eaten in 'epidemic' proportions that first caught my attention. What struck me most was that it seemed still to be working months or even years after being commenced. But was this so?

Late one autumn I chanced to see three patients at the City of London Migraine Clinic who gave me the very same story that I came to know so well: a long history of migraine from childhood; unresponsive to conventional preventive drugs; thought they would give feverfew a try, not believing for a minute that it would work – yet it did. Their last hope had succeeded where all else had failed.

I remember being impressed by the beneficial changes in the life styles of these three patients and humbled by the fact that it was not my treatment that had brought about these improvements. They were so bright and happy and seemed grateful to me, despite the fact that they had apparently found their own salvation. Also, I distinctly recall that after seeing the first patient my natural scepticism had me wondering, rather uncharitably, whether she might be a little eccentric. After all, who in their right minds would eat garden weeds? The answer came soon enough: those who were desperate.

2

The Welsh Connection

The suggestion that feverfew might be of value in the prevention of migraine came in 1978 and 1979 when several articles appeared in the national and provincial press.

An investigatory article in the health magazine *Prevention* reported the 'simple success story of a little weed' which had the power to cure classical migraine completely and, as a bonus relieve, if not reverse, advanced arthritic conditions. The article recounted the story of Mrs Ann Jenkins, a 68-year-old Cardiff doctor's wife, who had suffered migraine from the age of sixteen. An unusual feature of her migraine attacks was that they increased in frequency in middle age, so that by the age of fifty they were occurring at least every ten days and lasted for two or three days. She tried treatment after treatment, but to no avail. Medicines containing ergotamine made her vomit.

In 1973 a friend of Mrs Jenkins' sister mentioned Mrs Jenkins' plight to her elderly father. He had found feverfew effective in treating arthritis and suggested to Mrs Jenkins that it might be useful for migraine also. He sent her a plant with the advice that every day she should take a 'pinch' of fresh leaf.

In 1974 Mrs Jenkins decided to take a whole small leaf every day, 'chopped, in a bread and butter sandwich'. When nothing happened she increased the dose to three small leaves. Day after day she persevered and very little happened for five months when, despite the headaches persisting, the vomiting associated with the migraine stopped. She also found that she was using her Medihaler ergotamine less and less. After six months she went for a whole month without a headache. The occasional 'breakthrough' headache occurred until the tenth month, after which time she had no further attacks.

She told no one that she was taking the plant until she had convinced herself that it was helping; then she gave the plant to equally desperate migraine sufferers, some fifteen in all by 1978, and with similar results.

Mrs Jenkins' story is typical of the migraine-sufferer who

turns to feverfew. Her beneficial response and with no side-effects, after so many years of unsatisfactory orthodox treatment, led her to advocate it enthusiastically to others in the same predicament.

A summary of the account in *Prevention* was written up for the Sunday Express by Robert Chapman, Science correspondent, and published on May 21st 1978. For this account, the views of the London based Migraine Trust were sought and the then Director of the Trust, Derek Mullis, was quoted as saying:

'We are always hearing of unusual treatments for migraine, but few seem to remain effective for long. The use of feverfew is certainly interesting. I shall be making further inquiries about it.'

He did. He got in touch with me at King's College and we discussed the possibility of looking into the chemical constituents of the plant. For reasons given later in this chapter, I decided to do nothing for a few months. But meanwhile Mr Mullis passed on to me any interesting correspondence received by the Migraine Trust from feverfew users.

The articles about Mrs Jenkins had also mentioned the alleviation of her symptoms of arthritis, for which purpose feverfew was used as long ago as the sixteenth century. According to John Gerarde, the writer of *The Herball* or *General Historie of Plantes* in 1597, it was 'profitable applied to Saint Antonies fire, to all inflammation and hot swellings'.

Publicity for Mrs Jenkins' 'discovery' continued through to the late summer of 1978 with a television interview and a repeat of her story by Grafton Radcliffe of the *Glamorgan Gazette* in August. This coverage stimulated a great deal of interest among migraine and arthritis sufferers and hundreds of people contacted Mrs Jenkins for further information or with requests for feverfew plants or seeds. Mrs Jenkins produced a leaflet showing the amount in small or large leaves she took each day and giving general information on its cultivation. Hundreds of sufferers followed her example; many claimed benefit and were truly grateful.

The Migraine Trust was occasionally chided for not acknowledging the importance of this discovery, as by this correspondent in August 1979:

I have now been taking feverfew for one year, and it is like a miracle to me . . . I have many friends who have also benefitted and who feel as I do that a very big 'thank you' is due to Mrs Jenkins – and why not through the Migraine Trust.

Another sufferer who was greatly helped by Mrs Jenkins' treatment regimen wrote to the Trust:

I wish someone could do some research into this plant to see if it could, in the long run have any side-effects. As far as I am concerned now, I could not be more grateful for the valuable information Mrs Jenkins has given. Once again after more than twenty years I can lead a normal life.

In the media the news about feverfew was all good. In *Farmers Weekly*, Paul Lees wrote an enthusiastic account about a retired farmer, Herbert Swann of Great Totham, Essex, who, although not a migraine sufferer himself, grew the plant and dispatched it both fresh and dried all over the world. Mr Swann and his wife Christine had previously featured in the newsletter of the British Migraine Association, after which he was inundated by letters requesting plants. Professional herb growers, however, were not quick to respond to the demand.

The article in *Farmers Weekly* carried a statement by the then Hon. Secretary of the British Migraine Association, Peter Wilson, who said that the Association had been particularly interested in feverfew but that one or two people had experienced difficulties with it:

'We have had a few complaints that the user has reacted with blisters in the mouth and throat.'

Mr Wilson pointed out that the Association was interested in the possible testing of another herbal remedy for migraine, devil's claw, but that

' . . . we don't take these remedies too seriously in view of the highly likely placebo effect.'

These comments illustrate the dilemma in which migraine charities found themselves. While the public were reading about the advantages of feverfew, the charities were also receiving many letters from those whom the plant did not suit:

FEVERFEW

Birchington, Kent

I only took feverfew for about two weeks. Ulcers began to form in my mouth and I was afraid of being ill . . .

Manchester

She only took feverfew for about six weeks and then had a very intense allergic reaction to it – her mouth and lips swelled as though stung. She has not taken any since.

Deal, Kent

. . . She never received any benefit whatsoever from taking the feverfew, and eventually gave the whole thing a miss saying it was rubbish and that she had no confidence whatsoever in its so-called medicinal properties.

London, NW1

In fact, over the four months the feverfew did not make any difference.

Newport, Gwent

We were disappointed when the feverfew sandwiches did not help.

Totton, Southampton

However after taking half the suggested amount of two teaspoonsful of the granules (chrysanthemum and sugar crystals, sold as feverfew) dissolved in water, I experienced severe heart 'pumping' which was both distressing and frightening.

Colne, Lancashire

After about six migraine-free weeks my mouth and throat became so severely swollen that I could hardly manage to eat or drink . . .

So the charities could hardly be blamed for not joining with the popular press in extolling the virtues of feverfew.

Then came a further development.

In June, 1979, a surgeon writing in *Woman's Weekly* under the pen-name 'Doctor Margaret', alleged that feverfew could damage the liver. Readers had asked whether there were any dangers in using herbal remedies rather than artificially-

22

produced drugs, and Doctor Margaret replied:

> Yes, herbal remedies can, in some cases, be dangerous. They are not pure substances and most of them have not been completely analysed. Some – feverfew, for instance – contain quite potent liver poisons, which have caused illness, sometimes very serious.
>
> If you are taking other medicines, those may react disastrously with some of the herbal constituents. It is true that herbal treatment is satisfactory in a number of cases, but self-medication can be dangerous and you should ask your doctor whether he is happy for you to use herbs before you dose yourself with them.

As a medical practitioner I, too, tend to err on the side of caution and, when giving advice, try to present a balanced view. But, in a single sentence, feverfew was damned by Doctor Margaret as 'a potent liver poison'. As a matter of fact, little was known about the effects of feverfew on liver function in June 1979 and it was rather a strange example to use when there were so many other herbs *known* to harm the liver which would have served Doctor Margaret's purpose. Feverfew had, of course, been the subject of articles in the national and provincial press, but why had she linked it with liver damage? Perhaps she had confused it with another popular herb, comfrey, that had been the cause of some anxiety a few months earlier, as described in *The Guardian* on October 2nd 1978:

> Comfrey has joined the ranks of those things which modern research has discovered are likely to do man harm.
>
> The announcement has caused some pain in itself because it has been made by the Henry Doubleday Research Association – a body hitherto very keen on the herb's promotion as a substitute for meat.
>
> The Association's director, Mr Laurence Hills, said at the weekend that 'the use of comfrey internally must cease'. This follows the discovery in Melbourne of a possible link between comfrey and liver diseases, including liver cancer.

23

Mrs Ann Jenkins and others wrote to Doctor Margaret seeking evidence for her statement and its source. This was not forthcoming. Mrs Jenkins mentioned in her letter to Doctor Margaret on June 6th 1979 that research was being undertaken on feverfew at King's College. This information subsequently appeared, paraphrased, in Doctor Margaret's replies to other critical correspondents:

At present, an investigation into feverfew is going on at King's College Hospital, and we all await the outcome with much interest.

The article had an immediate and alarming effect on feverfew eaters, many of whom stopped taking it. This proved useful to me, as I wanted to know what happened when people stopped taking the plant, but it wasn't an experience enjoyed by those who stopped. As far as I was able to ascertain, a little over half of those who were taking feverfew in June 1979, stopped doing so as a result of reading, or hearing of the article in *Woman's Weekly*.

Typical of the letters received by the Migraine Trust was that from a housewife in Matlock written in July 1979, soon after Doctor Margaret's article appeared:

About a year ago I was recommended the 'weed' feverfew as a cure for migraine. To my delight it really worked, reducing attacks from at least two a week to one now and again, and the attacks I did have were not as acute and easily dismissed.

However, I read in a magazine recently that this weed, if eaten, could cause liver damage so I have ceased eating the three leaves a day. Now my migraines have returned to two a week.

The tablets 'Migraleve' sometimes help me, but not to the same extent that eating the leaves of feverfew did. I wondered if you could clarify this situation for me and tell me if feverfew really does cause liver or any other damage.

Even doctors enquired on behalf of their patients.

Clearly such letters had to be answered. I wanted to be reassuring but on the other hand I just did not know whether the plant was harmful. So my standard reply admitted my lack of knowledge on feverfew's efficacy and safety, but also included a summary of what was known about side-effects and the opinion that, until more was known about it, feverfew could not be recommended.

My typical reply to a query from a doctor was therefore:

City of London Migraine Clinic
June 9th, 1981

Feverfew (Chrysanthemum parthenium) is used by many migraine and arthritis sufferers to prevent attacks. Its pharmacology and toxicology are largely unknown but liver function tests of the few patients we have assessed have been normal.

Adverse effects include severe mouth ulceration, palpitations, epigastric discomfort and fleeting muscle pains on discontinuation. Its long-term effects are unknown and we cannot therefore recommend its use at the present time.

This essentially remained my view until January 1983, when, as the liver function tests of feverfew users were consistently normal, I felt confident enough to start a few patients on the plant.

But by September 1978 it had become apparent that the eating of feverfew was widespread, and represented a potential public health problem if, by chance, it should turn out to be harmful. On October 28th 1978 the *British Medical Journal* published what was then known of the scientific evidence to support the use of feverfew in the prevention of recurrent attacks of migraine. What was known was incomplete and mostly wrong, but this small article showed the degree of awareness by doctors of the lay use of feverfew.

In my earlier discussions with Derek Mullis, Director of the Migraine Trust, I had explained that a research project based simply on anecdotal newspaper reports was not on. Nothing would be wasted, however, were we to identify a reliable source of plants just in case I changed my mind. Soon afterwards one of my students located a large quantity of them growing in the gardens of King's College's own Plant Sciences Department in South London.

A highly speculative project such as the testing for biological activity of feverfew leaves was low on my list of priorities. The Migraine Trust gave the college £50 to cover part of the cost, but I knew that two thousand times this amount would be the realistic sum for a research programme to identify the active compounds. The project was therefore given as an exercise to an undergraduate Medical student John (now Dr) Rose who had taken a year away from the traditional medical curriculum to read for a BSc degree in Pharmacology.

John undertook a most workmanlike project and although his studies were based on very crude extracts, they indicated that feverfew did indeed possess biological activity which might be worth following up. Subsequently, John wrote up his observations for the *Migraine News* in characteristically self-critical and cautious terms.

By coincidence the idea of a chemical extraction programme had occurred simultaneously to Dr Peter Hylands, a lecturer of pharmacognosy at Chelsea College. He had been following the newspaper reports of feverfew and as he had a promising prospective PhD student, Deborah (now Dr) Jessup, in need of funding, he applied to the migraine charities for financial support. This was not forthcoming, but Peter and I were put in touch with one another and when Deborah was awarded a postgraduate studentship by Chelsea College, the three of us formed a research partnership.

In Welsh literature

The treatment of severe headache and probably migraine with feverfew is not new. In 1772, John Hill MD in his book, *The Family Herbal*, wrote of feverfew: 'In the worst headache this herb exceeds whatever else is known.' What *is* new is its rediscovery and use as a preventive two centuries later. This has been entirely by the patients who have passed on the information from one to another following press accounts of Mrs Jenkins' observations.

It seems particularly appropriate, therefore, that the connection of feverfew and headache can be traced back through the centuries in Welsh literature and I am particularly grateful to Dr G. Wynne Griffith of Beaumaris for his translation from the Welsh of this Appendix to a book entitled *A Welsh Botanology, Part the First: A systematic Catalogue of the native plants of the Isle of Anglesey*, published in 1813 by Hugh Davies:

Wermod Wen: Pyrethrum Parthenium; Common Feverfew
This common, innocent herb possesses medicinal properties beyond praise, and of great value in relation to women: there is no drug used by physicians as effective in easing and bringing on the periods, nor for all kinds of maternal disorders. It also is beneficial for the bowels, against worms, and the intermittent fever. Having made a strong infusion of the leaves with boiling water, the patient should drink about half a pint twice or thrice a day. A virtuous gentlewoman who had suffered greatly for a long time, so much so that in her youth she was driven almost insane at times with terrible headaches, frequently recurring, and having spent much money on physicians, to no use or profit whatsoever, in time was completely cured by this means: she drank copiously of the infusion of the leaves, and made a plaster or warm compress of the leaves from which the infusion had been made, to place on that part of the head where the headache was: the disease left her never to return.

The therapeutic value of feverfew has remained dormant possibly because aspirin and other similar drugs rendered it obsolete. Its use for headache was never widespread – it is hardly mentioned in the herbals for instance – and probably only became noticed again because of the poor efficacy and high incidence of side-effects associated with modern drug treatments of migraine.

3

The Alien

Feverfew is not indigenous to Britain. It was probably brought to this country for medicinal purposes by the Romans, although some authorities set the date later, in Anglo-Saxon times. It came to be grown in most gardens, from which it escaped to the hedgerows and dry stone walls which suit it so well. It was similarly introduced into North America as a medicinal plant by the early colonists.

Feverfew's Latin name is *Tanacetum parthenium* (syn: *Chrysanthemum parthenium*), and according to the Greek writer Plutarch, it acquired the name parthenium because it was used to save the life of a person who fell off the Parthenon during its construction in the fifth century BC. Its common English name, feverfew or feverfue, was possibly inspired by the feather-like leaves, although it is more likely to be derived from its Saxon name feferfuge or feferfugia, meaning that which dispels a fever. Certainly, it was used for this purpose.

Alternative medieval Latin names included Amaracus, Matricaria and Amarella. The English also knew it as whitewurte and St Peter's wurt, the French as espargoutte or matricaire, and the Germans as mutterkraut.

Confusion over names

Although feverfew is an easy plant to identify, some people find its various botanical names confusing. In February, 1983 a Chelmsford man wrote to me:

> Having recently read a letter about feverfew in *Migraine News* 44, I wrote to the Migraine Trust asking for clarification about the naming of plants commonly known as feverfew. There seems to be a variety of synonyms . . . Among the names found in seed catalogues and elsewhere are *Pyrethrum parthenium*, *Pyrethrum matricaria*, *Matricaria*, *Chrysanthemum parthenium* and *Tanacetum*. There are also coloured

and double forms . . . *Chrysanthemum parthenium* appears
to be the most acceptable botanical name. I am anxious to
grow feverfew both as an ornamental garden plant and as a
possible remedy for migraine. I should be grateful for any
comments you could make to assist me in sorting out this
tangle.

In fact only one name is correct botanically for any plant at any
one time. But plant names are changed from time to time and
feverfew was recently reclassified from the *Chrysanthemum* to
the *Tanacetum* (tansy) genus. Once a name has come to be
widely accepted and used, it takes a long time to establish the
new, more correct name – and this is the case with feverfew.
There is no longer any justification for ascribing it to the Chamo-
mile genus *Matricaria*, or to the genus *Pyrethrum*.

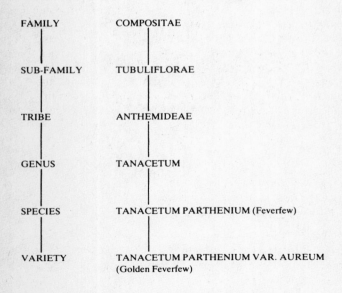

FAMILY	COMPOSITAE
SUB-FAMILY	TUBULIFLORAE
TRIBE	ANTHEMIDEAE
GENUS	TANACETUM
SPECIES	TANACETUM PARTHENIUM (Feverfew)
VARIETY	TANACETUM PARTHENIUM VAR. AUREUM (Golden Feverfew)

The botanical classification of Feverfew

Varieties of plant species are correctly written with the syllable var. appearing between the species name and variety name. Thus the golden variety of feverfew is *Tanacetum parthenium var. aureum* but often the var. is omitted.

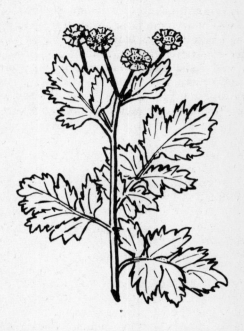

FEVERFEW: *Tanacetum parthenium* [(L.) Schultz Bip.]

Where to find feverfew

Common feverfew is very different from other plants bearing the name feverfew, such as sea-feverfew, corn feverfew, and sweet feverfew. It grows in most parts of the British Isles and is often found growing wild in old gardens and on waste land, where it thrives almost as well as when cultivated.

It will, in fact, grow almost anywhere, and certainly seems to enjoy rooting between the stones of walls such as those found in Britain in Devon, North Wales and Yorkshire. It grows throughout Europe, North Africa, China, Japan and Australasia. It is not averse to shade, but tends to grow most prolifically in the sun.

Many people all over the world are now growing their own feverfew, not only to provide sufficient leaves for consumption throughout the year, but also as an ornamental plant for the herbaceous border. It is a perennial growing to a height of 14 to 45 cm with yellow-green leaves, having deeply cut segments. The leaves have an extremely strong and characteristic aromatic odour when crushed. They are also bitter to taste. The attractive daisy-like flowers, which can measure almost 2 cm across in the right growing environment, form loose flat-topped clusters. The flowers have approximately twenty central yellow florets formed into a disc with short, broad, white florets radiating from the outer margin of the disc. The florets give rise to flattened, ribbed seeds about 1 to 2 mm in length.

The feverfew flowers from late June to October, but the long, woody, flower-bearing stems provide fewer leaves suitable for medicinal use. For this reason, many users nip out the flower buds of three-quarters of their plants, allowing only a few to go on to provide seed for propagation purposes.

Ornamental feverfews

Alternative feverfews for ornamental purposes, especially for the front of a border, are the golden varieties. These are smaller plants than the common feverfew, having beautiful golden leaves which hug the ground, and little button flowers that bloom from July to October. They do well in a sunny position and, like the common variety, remain in leaf throughout winter.

The varieties Gold Star and Golden Ball both grow to a height of about 20 cm. Double-petalled white cultivated varieties, Silver Ball and White Star grow to the same height as common feverfew. They make superb edging plants for a bed of Gold Star. The leaves of the cultivated varieties contain chemicals similar to those of common feverfew and are probably similarly effective when used medicinally, although this has to be verified. The cultivated varieties are often the ones to be found on sale at garden centres and nurseries.

How to grow and propagate feverfew

Choose a sunny bed containing well-drained soil and dig in compost or rotted grass cuttings. There is no need to worry about removing any stones. Seedlings can be planted out any time if transferred from an outside bed. They should be planted about 2 ft apart, unless they are being grown to provide cuttings, when they can be planted 1 ft apart, as alternative plants will be removed eventually. To ensure a good supply of leaves each person taking feverfew needs at least three and preferably four or five plants.

Feverfew seeds usually germinate well, provided they are less than one year old. Their germination is considerably reduced if they are allowed to become too dry for long periods. Sow the seeds in ½-inch deep rows in trays containing garden soil or a multipurpose compost and pot out into 2-inch diameter pots when the seedlings are 1 inch high. The roots are very well formed and even on the smaller seedling can be 2 to 3 inches in length. The plants can either be 'potted on' in 5 to 6 inch pots and kept in a cold frame, or planted out into beds when they are 2 to 3 inches in diameter. They are extremely hardy.

Feverfew will also grow indoors and this is especially useful in winter. The plants require plenty of light, warmth, and daily watering to ensure a lush growth of leaves.

Feverfew can be propagated from cuttings taken from a plant of any age, but it does particularly well if taken from one year old stock after flowering, when the woody stems have died back. Roots grow down from the junction between a new shoot and the old stem. But the most reliable form of propagation is by root cuttings.

First lift the plant and shake off the soil, then, using an old razor blade, slit the main stem along its mid-line upwards through the leaves, and downwards through the main root. The resultant two halves can then be planted out in beds or pots. Feverfew plants can also be propagated from leaves whose stems are dipped in hormone rooting powder, but the failure rate is high unless done by an expert. Sprigs of new growth placed in wet soil develop roots 1 to 2 inches in length within fourteen days.

Pests and disease

In a wet year feverfew is susceptible to mildew and young plants may well 'damp off'. Black fly can be troublesome on new shoots in the late spring but can be washed off with a hose. As with chrysanthemum, when the plants are in flower, watch out for earwigs in wet weather. Ladybirds and white spiders seem to be particularly attracted to the plant especially when the seed heads are ripening.

Where to obtain feverfew plants

Many thousands of people world-wide now grow the plant for its medicinal properties and are usually only too keen to pass on seedlings. If you don't know anyone who grows feverfew, ask at your local garden centre or herb nursery, or in the UK write to:

>Ashfields Herb Nursery,
>Hinstock, Market Drayton, Shropshire.
>tel: 095 279 392

>Stubbings Country Herbs, Stubbings House,
>Henley Road, Maidenhead, SL6 6QL
>tel: 062882 5454

The latter guarantees to provide the wild variety and have produced a useful pamphlet.

Seeds can be obtained from:

>Messrs. Thomson & Morgan,
>London Road, Ipswich, Suffolk
>tel: 0473 218821

If in doubt

The physical appearance of feverfew is so distinct from the related chrysanthemums, tansies and chamomiles that its identification should not be a problem. I have encountered only one person out of 300 surveyed who was not actually eating feverfew – though she thought she was. I have also encountered twenty other people who were not absolutely certain as to what they were taking, although in every case it turned out to be feverfew.

Nevertheless, if you are in doubt as to the identity of a plant, check first with your local nurseryman, horticultural society or the Botany Department of the nearest university. Otherwise, consult one of the floras in the reference section of your local public library such as the *Flora Europaea*, 1976, Volume 4, published by the Cambridge University Press. But remember, plants can be harmful; be sure before you try and if in doubt, don't! In any case, anyone wishing to try feverfew should discuss this first with their doctor.

Feverfew on holiday and in winter

When migraine sufferers stop taking feverfew after several months of daily use their attacks quickly return. This can be so disabling that some people restrict their movements to within a day's journey from their source of supply. These individuals, although relatively few in number, never take holidays and when suddenly and unexpectedly faced with the prospect of journeying to foreign parts, become stricken with panic and refuse to go. They have not heard of the alternative means of taking feverfew (see pages 35 to 40).

Winter, too, brings its problems as plant growth may be severely retarded at this time and not everybody is able to pot up the plants and take them indoors. A 71-year-old retired Major from south-west London wrote:

> In recent springs and summers I have had great relief from eating a morning sandwich of a tablespoonful of feverfew leaves in bread but face a long winter without abatement of migraine almost three or four times each week because the

plant does not show leaf again until January/February (if then). I have continuously asked all over the country why no tablet or 'infusion' is available from 'Health' shops – the usual reaction is: 'never heard of it!' You will be doing many thousands of sufferers a great service if you can persuade the Ministry or in some way promote the provision of feverfew in some other form than the leaves taken fresh from the plant, so that it can be taken in liquid or tablet form during the 'fall' and winter. I can assure you that it works – without it I am entirely dependent on diet and *Migril* which I have taken for thirty years.

In fact, advertisements for dried feverfew capsules or powder do appear in health magazines and retail outlets in the UK. At the time of writing, capsules and tablets containing high doses of feverfew (100 mg or over), and those whose dose is not speci- fied, have not yet been properly assessed for safety and efficacy and are better avoided. There is no convincing evidence that homoeopathic feverfew preparations are effective in migraine either; indeed the use of a dilution of feverfew runs counter to basic homoeopathic principles. And of some twenty patients I have encountered using the 6X homoeopathic potency feverfew tablets now on sale, only one person thought they might be of benefit.

Recently a tablet containing 25 mg dried feverfew has been marketed by an established manufacturer in the 'health food' business. This strength does permit some adjustment of dose to be made without the danger of overdosage.

Capsules and tablets mask the bitter taste of the plant and are readily transportable when the user takes a holiday. They also serve as an ideal stop-gap when the leaves are frosted or snow covered during winter. But, despite their convenience, they are by no means the only alternative way of taking feverfew.

Granules of 'chrysanthemum' powder mixed with cane sugar and imported from China is marketed widely in 'health' shops as feverfew. I know of many who have tried a sachet or teas- poonful each day as an infusion and have found it of benefit, although others have experienced palpitations and a rapid pulse rate. As there is some doubt as to whether the contents really are *Tanacetum parthenium* – the labelling is in Chinese – it is probably wiser not to take large amounts of this formulation every day.

Air-dried leaves and powdered feverfew leaves can be obtained from qualified herbalists and some herb suppliers. If you pack the leaves separately between layers of kitchen film or foil the daily dose can be measured with accuracy. Once dried leaves are crushed however, the daily quantities are difficult to measure out, but according to Mrs Jenkins powdered feverfew equivalent to a large leaf just fills the heaped bowl of an ordinary salt spoon.

Many users have solved the supply problem in other ways. Some pick long, leaf-bearing stalks and insert them into round-based bottles with rubber seals at the top, the sort used to transport single orchids through the post. This works quite well but if the leaves are kept under water they will rot and give off a putrid odour.

Mrs Lesley Hirst of Kent has discovered an ingeniously simple way to prepare what she calls Dried feverfew 'tablets' and sends out the following instructions to people who ask her for feverfew plants:

Dried feverfew tablets

Dry 100 feverfew leaves on a sheet of kitchen roll in your airing cupboard. When completely dry crush the leaves as small as possible (an electric liquidizer is ideal). Mix the crushed leaves with 3 oz icing sugar and approximately 1½ teaspoons of cold water to form a very stiff dough that can easily be handled. Then pick off small pieces of the mixture and roll them into pea-sized balls. Dredge a baking tray with icing sugar and place the wet 'tablets' on this. Place in an airing cupboard for about 24 hours to dry the tablets thoroughly, and then store in an airtight container. You should be able to make 100 tablets from this recipe, therefore take two tablets daily in place of the fresh leaves.

Although Mrs Hirst's recipe will produce pills containing

varying amounts of feverfew, they probably do not vary in weight much more than raw leaves do from day to day. Furthermore, she has taken the important step of producing a 'standard leaf' unit and in this way protects herself from accidental overdose – which can easily occur with the dried powder. An early sign of overdose is abdominal pain and purging which may last for a couple of days. One lady in her forties who had never taken feverfew before took a heaped teaspoonful of finely powdered feverfew just to see what it would taste like. She found the taste quite acceptable – 'herby' – but not the colicky diarrhoea that followed.

Drying feverfew

Users find that dried feverfew seems to work just as well as the raw leaves. When patients at the City of London Migraine Clinic in England were changed over to capsules containing the equivalent in powder to their daily dose of leaves they noticed no difference between the two formulations.

A lady from the North of England who suffered from disabling classical migraine visited me at the clinic in May 1981. She had been taking feverfew since 1978 and had had no more attacks. She was greatly worried about a forthcoming visit to Australia and New Zealand that was to take her away from England for ten months. I advised her to dry sufficient feverfew during the next three months for the duration of her trip. On her return she wrote me the following progress report:

> Further to my visit to you on May 26th 1981, I have to report the following progress with the use of *dried* feverfew during my absence from England between August 11th 1981 and June 19th 1982 . . .
>
> On Dec. 2nd 1981 I experienced the first stage of a migraine – visual disturbance – but it went no further.
>
> On May 10th 1982 I had the visual disturbance followed by another lot on May 15th. Neither went beyond the first stage.
>
> We returned to England on June 19th when I resumed taking fresh feverfew but I had another bout of visual disturbance on July 4th and again on August 2nd. Once more

there was no development beyond the first stage.

Since then I have been free of any further attacks – and hope to continue so!

Whilst in New Zealand I gave some dried feverfew to a friend who has suffered with migraine for many years. The effect was instantaneous, unlike the time lapse of three months I experienced, and she has been clear of any attacks since first taking feverfew in December 1981.

Although more than eighty per cent of users still eat raw leaves, an increasing number are taking to the dried form.

A little folklore appears to be building up around the drying procedure to the effect that the leaves should be picked in the morning on a sunny day when the dew has dried and shouldn't be washed and so on. Take my advice! Pick them at any time of day, wet or dry, and wash thoroughly under cold water. Spin off the excess water in a salad drier or shake vigorously in a colander, and then dry on blotting paper, or kitchen roll in an airing cupboard or a dry well-ventilated room. To ensure even drying, turn the leaves at intervals.

The leaves should be completely dried; otherwise they will attract mildew and be unusable. As a guide, completely dry leaves weigh approximately one sixth of their wet weight.

Feverfew keeps well when frozen and can be stored in the freezer until required, when it can be thawed and used immediately. Freezing helps to break up the cells and release the active contents. The frozen leaves may be dried like fresh ones and can be powdered before or after the drying process.

Store the dried leaves in an air-tight container such as a glass Kilner jar. Ensure complete dryness in the storage jar by including a muslin bag containing a suitable drying agent such as silica gel, obtainable from most pharmacists.

Masking the taste

Most users cannot stand the extremely bitter taste, reminiscent of quinine, and some give up for this reason. Most of those who persist have found ways of masking the taste. A popular method, and advocated by Mrs Ann Jenkins, is to eat the leaves in a bread and butter sandwich to which some honey or sugar has been added. Honey was the original method of disguising the taste recommended in the herbals. Others chop up the leaves and sprinkle them over their lunch or dinner as a garnish, rather like chopped parsley. Mrs Hirst, whose recipe for Dried Feverfew Tablets appears on page 36, sends out the following advice:

If, like me, you find the leaves utterly disgusting to eat, try the following method:

Take a jelly cube and, using a sharp knife, cut a hole in the centre. Turn the cube on its side so that you can see how deep the hole is, then cut across the cube just below the hole. Turn the cube upright again and trim off the corners to a 'swallow-able' size. Roll up two feverfew leaves and stuff them into the hole. The cube is then quite palatable to swallow with a drink.

An arthritis sufferer from Somerset who claims benefit from daily doses of feverfew disguises the taste by chewing two leaves with a date. A Birmingham woman with rheumatoid arthritis takes her dose with 'crispbread and cottage cheese'. Others mix it with chopped parsley.

How it is taken

Most feverfew users eat two or three of the raw leaves every day as a preventive for either migraine or arthritis.

Some people use feverfew in much larger doses to ward off a migraine attack as soon as they get the first warning signs. A heaped tablespoonful is not unusual and it seems to work. An interesting variant is that of a woman in Stoke-on-Trent, who wrote to me in February, 1983, about a migraine attack which occurred as she was buying vegetables from a local allotment holder:

It was a very sunny day and I could have cried with the pain, it was so bad . . .

The allotment holder told me he had just the thing to give

me relief. He then gave me the leaves of the plant and told me to chew them for a little while, swallowing the juices but not the leaves.

I was a little wary at first as the taste was very bitter. But the relief was unbelievable.

Treatment of the acute attack is how feverfew was used traditionally for inflammatory disorders. It was given in the single high dose of 2 drams of powder taken with honey, which is equivalent to $1/8$ oz. This is about the weight of 140 dried small leaves measuring $1\frac{1}{2}$ inch $\times$ $1\frac{1}{4}$ inch. Thus the single dose for treatment of acute attacks recommended by the writers of the herbals is equivalent to more than the average amount consumed over two months when used as a preventive. The latter use is a recent innovation largely pioneered by Mrs Ann Jenkins.

Even larger amounts, such as 1 oz of leaves and flowers, were used for the purpose of making an infusion or tea, and are still used by some people. The 'tea', that today is often sold by 'health' shops as feverfew, is the chinese chrysanthemum mixed with cane sugar crystals. Nowhere on the pack is feverfew mentioned but it does seem to help some people, as witnessed in January, 1983, by the wife of a Scottish clergyman who suffered from migraine.

My husband has, in the last two years, developed acute migraine averaging two attacks per week . . . However, in despair we decided to try chrysanthemum tea which we bought as granules from a health shop. The result has been miraculous. He has not had a headache for over a month now.

Some people make their own tea – Mrs A of Coventry, for example:

I have used it for some years. We grow the herb in our garden. I pluck the stems, dry them and break down the leaves, which I store as tea. When needed, I add boiling water, about 1 pint to a couple of teaspoons, infuse for 24 hours, strain and add a little of the liquid to my morning drink. Not a very strong dose, I know, but it seems to prevent my having the regular attacks which were quite severe – three days in bed with headache, nausea and sometimes vomiting.

Now I have a very slight attack infrequently, don't have to stay in bed, suffer no nausea and am able to eat normally . . .

Obviously, the popularity of feverfew as a treatment was growing rapidly, but major questions remained unanswered. Did it really work? Could evidence be acquired that would satisfy the most sceptical of medical scientists? Was it working by suggestion in severely afflicted people willing to grasp any straw? Was it safe to take? It seemed to me that the first step was to contact as many feverfew users as possible and to discover what proportion had apparently gained some benefit from long-term use.

4

Does Feverfew Work?

It was obvious from the beginning that there would be no point in expending a great deal of effort and money on laboratory testing unless I had more reliable information on the effectiveness of feverfew in migraine than merely anecdotal press reports. So my wife and I compiled a comprehensive questionnaire to be distributed to as many long-term feverfew users as we could find. It was to be restricted to those taking the plant for the prevention of migraine, and included questions on the parts of the plant eaten, how much, how it is taken and for how long, the side-effects encountered and what happened when individuals stopped taking it. The questionnaire is reproduced at the end of this book.

The day after we compiled the questionnaire I received a telephone call from a general practitioner in Bearsted, Kent, Dr John Gledhill. He had undertaken an open study of feverfew leaves on his own patients and felt certain there were grounds for believing that it worked. He had contacted a major pharmaceutical company and had been interviewed by one of its scientists, but had heard no more. I told Dr Gledhill of the questionnaire and asked if he would distribute it to his patients. He agreed and I sent him thirty copies, most of which were duly completed and returned.

Dr Gledhill's was the first clinical appraisal of feverfew that I had come across and it encouraged me to continue. What is more, a major drug company had considered it worthwhile sending one of its senior personnel 250 miles to investigate.

The feverfew users who were my patients at the City Clinic and those provided by Dr Gledhill gave me forty-five to fifty replies. I had only been thinking in terms of twenty or so anyway, so was well-pleased. Then I remembered the offer made by Mrs Jenkins to the Migraine Trust in a letter dated November 2nd 1978:

Should any independent authoritative body wish to undertake an epidemiological survey on the use and value of this

42

herb, I would, with the consent of the persons concerned be ready to supply names and addresses.

I wrote to her and she sent me over 700 names and addresses, for which I shall always be indebted to her.

Of all the people contacted, from whatever source, we heard from 489 – that is about fifty-five per cent. Of these, 195 had taken feverfew for arthritis or had not taken it at all; some had originally enquired on behalf of others; thirteen had died in the meantime. Eventually we had about 200 completed forms, and then in due course others heard about the questionnaire, or wrote to the clinic about their own experiences with feverfew so that another hundred completed forms were eventually received.

Forms were returned by individuals living as far apart as Penzance in south-west Cornwall and Elgin in the north of Scotland. As can be seen from the map of the British Isles, most long-term feverfew users with whom contact was made lived in London, Manchester, Merseyside, the Midlands and South Wales. The cluster in Kent in south-east England were Dr Gledhill's contacts.

Characteristics of feverfew-users

The ratio of women to men users was 5.3 to 1, their average ages were fifty-five and fifty-four years respectively, and the information they provided showed that eighty-eight per cent were suffering true migraine attacks. A further eight per cent suffered from daily (tension) headache prior to feverfew-taking.

In fifty per cent the attacks began before they reached twenty years of age and in seventy-five per cent by the age of thirty. Their trigger factors resembled those reported in other surveys of migraine sufferers (see page 8). In fact, these feverfew users resembled other migraine patients in every respect but one – their headaches tended not to respond to conventional medication.

Nearly eighty per cent said they had learned of the plant from reports in newspapers or magazines. This underlines the influ-

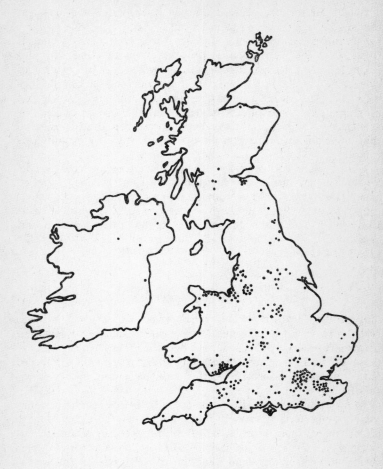

Homes of 300 Feverfew users in the British Isles

ence of the popular press – nearly all of these individuals were prepared to take the plant every day without seeking medical advice as to the wisdom of it. And up to that time, no newspaper or magazine had suggested that the plant might be other than a harmless 'natural' cure, free of unwanted side-effects, the implication being that feverfew was a herb, not a drug – and only drugs had side-effects. (The fact is, however, that once a herb is used as a medicine it becomes, by definition, a drug.)

Only one in eight users did not start taking feverfew as a result of reading about it in the press. Among this group were individuals who learned of the plant from personal contacts. One patient of mine first heard of feverfew from a young girl whose grandfather had eaten it every day to prevent his headaches from recurring. This is one of the very few instances I have come across in which feverfew had been used as a treatment for migraine before the publicity given to Mrs Jenkins' observations in the late seventies. It indicates that this treatment of migraine might have been known to folk medicine.

The source and varieties

The majority of users (eighty-eight per cent) grew their own feverfew plants or obtained them from the garden of a friend or relative. There was considerable variation in the shape of the leaves consumed and it was obvious that several varieties were being used. Some of the different leaf shapes are shown on page 47. The shapes of feverfew leaves may be different for plants of different ages as well as between different varieties. They all have the characteristic odour of feverfew which, as far as the medicinal properties of the plant are concerned, possibly has more significance than the shape of the leaves.

Leaf colour is not an accurate indication of the variety of feverfew either. Root cuttings of wild feverfew planted out in soil rapidly yellow, whereas those from the same plants, simultaneously kept in a cold greenhouse in permanently wet soil, remain green.

It is likely that the majority of garden centres and nurserymen sell the cultivated varieties but, at the time of writing, those mentioned in Chapter 3 stock the common wild variety.

What part of the plant is eaten?

Ninety-two per cent of all users eat the leaves, seven per cent also eat the soft stems of new growth with the leaves. If the plant is allowed to bolt into flower it produces far more stem than leaves, so nip out the flowering shoots early to encourage leaf growth. However, I have come across five users who eat not only the leaves and stems but flowers too, when in bloom.

The vast majority of feverfew-users eat the plant freshly-picked although many long-term users have found that the dried leaf works just as well, and as discussed in Chapter 3, the dried leaf is more convenient for holiday and winter use.

Five out of six people take the leaves with food, water or both. Of those who take it without the assistance of these, some actually like the bitter flavour.

As a rule, feverfew is taken once every day but there is no preferred time. The 'dose' taken varies considerably, because of the marked variability in the size and shape of leaves of different varieties of plants and plants of different ages. When the leaf dose was standardized into 'small leaf units' measuring 1½ inches long by 1¼ inches wide, it was found that the average consumption was just over two-and-a-half leaves every day.

Are feverfew-eaters more likely to use herbs than conventional medicines?

The answer to this is a decisive 'no'. Seventy-five per cent of all users had never taken any other herbal remedies, and of those who had, many had taken vitamin preparations bought over the counter, which couldn't really be classed as herbal products. In fact, nearly all patients had taken medicines prescribed by their family doctors. For example, seventy-five per cent had taken ergotamine at some time or other, indicating that they were so severely afflicted they had been to their doctor for prescription medicines. Usually, ergotamine is only prescribed when patients fail to respond to simple pain-killers.

Fewer than half (about forty-five per cent), had found the common pain-killers aspirin and paracetamol to be of any help. However two other treatments used for treating the symptoms

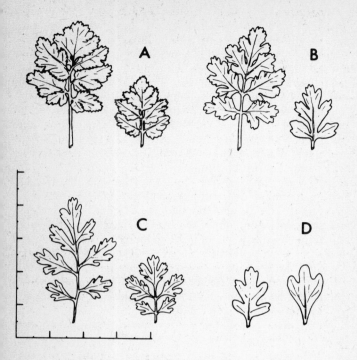

Drawings of feverfew leaves picked on the same day illustrating differences due to age and variety of plant. The scale is in inches.

A Common Feverfew: large and small leaves from a ten-week-old plant, before flowering.

B Common Feverfew: large and small leaves from the base of an eight-month-old plant, in flower. Note the loss of the feathery outline that was a feature of the immature plant.

C Cultivated Feverfew, variety Silver Ball: large and small leaves from the base of an eight-month-old plant, in flower. Note the greater separation of the leaflets which never possess a feathery outline at any stage. The young and mature basal leaves of this variety are identical in shape.

D Small leaves from the tops of flowering stems; left, common feverfew; right, cultivated feverfew.

of migraine attacks had been relatively successful in a greater proportion of individuals. The first of these was Migraleve, available on prescription or over the counter at chemists. Migraleve combats nausea as well as being a pain-killer. Those who found Migraleve to be effective (see the table below), were also more likely to continue to use it. The second was ergotamine, which appeared to be equally as effective as Migraleve. However, when feverfew was taken, users tended to stop ergotamine twice as often as Migraleve. This almost certainly was because ergotamine is associated with a high incidence of side-effects unlike Migraleve; also, it cannot be obtained without a prescription.

Comparatively few people found preventive medicines, which had to be taken every day prior to the use of feverfew, to be of any help. Of those who had found them satisfactory, sixty-six per cent found them no longer necessary once they started taking feverfew. Clonidine (Dixarit), the preventive drug most commonly tried by feverfew users at some time or other, was found to be ineffective by seventy-two per cent. It is interesting to compare the success rates of other preventives listed in the following table with those achieved by feverfew.

The effectiveness reported by feverfew users of drugs commonly used in the treatment of migraine

DRUG	HELPFUL	NOT HELPFUL
For treatment of the acute attack:		
Aspirin	47.4	52.6
Paracetamol	44.4	55.6
Migraleve®	57.4	42.6
Ergotamine	61.7	38.3
Preventives:		
Clonidine	28.3	71.7
Methysergide	46.7	53.3
Pizotifen	53.8	46.2
Propranolol	47.1	52.9
(Feverfew)	(71.8)	(28.2)

The response to feverfew

Participants in the questionnaire were asked whether their headaches during use of feverfew were less frequent, more frequent; less painful, more painful – or unchanged.

Of the 300 individuals involved, twenty-three were suffering from daily tension headache and not migraine, a further twenty were taking conventional medicines used to prevent migraine attacks from occurring. This left for analysis 253 sets of results from patients who apparently suffered a true migraine. Of these, seventy-two per cent claimed that their headaches, were less frequent, less painful or both; twenty-six per cent thought they were unchanged, and two per cent considered that their headaches worsened while they were taking feverfew.

We also wanted to learn how other types of headache responded to feverfew, so the replies from the twenty-three people who were suffering from tension headache were analysed separately. Of this small group, seventy-eight per cent credited their reduction in headache frequency and severity to feverfew.

It is well-known in medicine that common conditions occur commonly together, so we felt it was particularly important to ask all of the feverfew-using migraine sufferers whether they suffered from any other illness. In fact, forty-seven per cent said they did and their illnesses are listed in the table overleaf. We also wanted to know whether the presence of another illness or its treatment might have influenced the high response rate, so my colleague from the King's College Computer Centre, Miss Margaret Skinner and I analysed separately the data for those migraine patients who were *not* suffering from any other condition or taking any treatment likely to confuse the evaluation of feverfew in combating migraine. In this group, seventy-two per cent claimed that their headaches were less frequent, less painful or both; twenty-four per cent thought they were unchanged and two per cent were made worse by taking feverfew. Thus the success rate reported by those not suffering from any other illness or taking any other treatments was virtually identical to the success rate of those with other conditions for which additional medicines were being taken. There is, therefore, no reason to believe that feverfew has been credited with successful responses that were actually brought about by other drugs that patients happened to be taking simultaneously.

**Medical condition(s) or symptom(s) (event)
suffered by feverfew users in addition to migraine**

	Number of Patients		Number of Patients
Allergy (including asthma, hay fever, angioneurotic oedema)	12	Intestinal hurry, irritable bowel	2
Anaemia	2	Mastoiditis	1
Angina	3	Menopausal symptoms	5
Atherosclerosis	2	Myxoedema (post thyroidectomy)	8
Anxiety	6	Osteoarthritis (including 'slipped disc', sciatica, back pain, cervical spondylosis)	11
Cataract	1		
Chronic bronchitis	2		
Cystitis	3		
Depression	7	Palpitations	2
Diabetes (mild)	1	Peptic ulcer	4
Diverticulitis, diverticulosis	5	Psoriasis	1
Epilepsy	1	Pyelonephritis ('kidney infection')	2
Fluid retention	1		
Gout	1	Rheumatoid arthritis	13
Hiatus hernia, indigestion with acid reflux, nervous stomach, nausea and vomiting	8	Sarcoidosis	1
		Sinusitis, catarrh	4
		Stomach abscess	1
		Thrush, vaginal infection	2
Hypertension	17	Ulcerative colitis, colitis	2
Insomnia	3	Vertigo	2

The second way in which we assessed the beneficial response rate to feverfew was to note down the actual numbers of attacks each month before and during the period on feverfew. This was done for 242 patients. It was found that thirty-three per cent no longer had any migraine attacks at all while taking feverfew and, overall, some seventy-six per cent had fewer migraines each month than they had before taking the leaves. Thus the two methods of assessing the effectiveness of feverfew on migraine headaches yielded almost identical results.

Many of those who completed the questionnaires were subsequently invited to attend the City of London Migraine Clinic, where their case histories were studied in more detail. Three interesting additional facts emerged. Those who, prior to taking feverfew, had suffered from nausea and vomiting during the headache phase, reported that they no longer vomited and seldom felt sick while on the plant. Secondly, when they did get a migraine it tended to respond more readily to simple painkillers and was less likely to incapacitate them. Thirdly, a proportion of those patients with classical migraine occasionally experienced the aura but no headache. Although this separation of migraine symptoms is not unknown to migraine specialists, it was interesting that feverfew seemed capable of influencing the course of an attack in this way.

The beneficial effects of most medicines often depend on the amount taken. For example, a patient given drug treatment for high blood pressure is often told that he might have to increase the dose if the response to the initial amount is inadequate. The correct quantity of feverfew leaves to induce an optimal inhibition of migraine attacks has never been determined. Nor has it been established whether one is more likely to get side-effects with five leaves than with one leaf a day.

The response to feverfew of migraine sufferers with other conditions

It was possible that migraine patients who also had another disease or illness might be less likely to respond to feverfew. So a separate analysis was made of those suffering from each of the six most common additional conditions. These included: allergic conditions such as asthma and hayfever; anxiety and depression; high blood pressure; osteoarthritic disorders including slipped disc and cervical spondylosis; a variety of stomach disorders such as peptic ulceration and hiatus hernia; rheumatoid arthritis. It can be seen from the following table that with the possible exception of allergic disorders, an additional illness did not seem to affect the response rate to feverfew.

Is the favourable reaction psychological

It is known that many illnesses seem to respond to suggestion, and this is called the placebo response. In effect, it is an im-

Proportion of migraine sufferers with other conditions claiming improvement in frequency, severity or duration of symptoms while taking feverfew

Condition	% improved	% with no attacks	% not improved	Number
Allergic disorders	58.3	16.7	41.7	12
Anxiety and depression	69.2	7.7	30.8	13
High blood pressure	76.5	29.4	23.5	17
Osteoarthritis	72.7	54.5	27.3	11
Stomach disorders	91.7	25.0	8.3	12
Rheumatoid arthritis	84.6	38.5	15.4	13
Overall	75.6	25.6	24.4	78

provement due to the fact that the patient gains relief from sharing his or her problem with the doctor. Usually the beneficial placebo response does not last for long and migraine attacks tend to return to their pretreatment levels after eight to twelve weeks. Patients discover this when they experiment for themselves with proprietary medicines; initially they seem to respond, but later they do not.

Now the patients who eventually take feverfew have usually tried everything else that is available, and have found that nothing works. These individuals are the least likely to respond to placebo as everything else has failed.

When I read about feverfew I really thought no more than 'Oh, well, I'll try anything once.'

They are also usually very surprised when they find that feverfew *has* helped them. Some are so anxious not to lose their new found freedom from migraine that they flatly refuse to participate in a clinical trial in case they are given a dummy (placebo) medication rather than feverfew and thus risk a recurrence of their attacks.

Two particular observations strongly suggest that the beneficial response to feverfew cannot be attributed to a placebo effect. First, the benefit lasts for as long as the individual takes the leaves – and the average duration of treatment is 2.3 years for men and 2.6 years for women. I know of no placebo effect that lasts for this length of time. Second, when a patient who has apparently responded to feverfew treatment suddenly stops taking the leaves the attacks recur with their former frequency and severity even, in some cases, after four or five years of continuous treatment.

It could be argued that the patient's knowledge that the treatment has changed somehow triggers the return of the migraines. This does not seem to be the case, however. When patients enter a clinical trial which involves them being taken off the leaves and instead being put on capsules containing their former daily amount of feverfew in freeze-dried powdered form or a placebo capsule identical in external appearance, only those taking the placebo suffer a recurrence of their migraine attacks. If the medication is then switched to the feverfew powder, the attacks again grow less. I have repeated this experiment several times, with the magnificent co-operation of a few patients interested in challenging the possibility that feverfew might be acting by suggestion.

A double-blind trial – that is, one in which neither doctor nor patient knows the identity of the treatment that the patient is receiving – has just been completed. The patients received capsules containing either feverfew or placebo. The results of this trial (completed only when this book was in proof) have confirmed that feverfew actually works as a migraine preventive.

Compatibility of feverfew and medicines

Many different drugs were taken simultaneously with feverfew, sometimes for long periods. There was no evidence that feverfew reacted adversely with these medicines, or that it neutralized their effects. Similarly the menstrual cycle was not upset in women taking both feverfew and oral contraceptive pills and there is no reason to believe that feverfew reduces the ability of the pill to protect against pregnancy.

Do the beneficial effects last?

The benefits certainly seem to last, but it is important to bear in mind that feverfew might only be suppressing the symptoms, not curing the condition. When individuals stop taking feverfew their migraines return, often within one week, and they are usually as severe as they were before treatment with the plant was started. If feverfew consumption is resumed the attacks disappear again. The curious fact is that in such cases they are controlled almost at once whereas when the same individuals initially started taking feverfew, several weeks or months usually passed before the attack frequency was reduced or the attacks disappeared altogether.

The University of London Computer picked out individuals who had taken feverfew daily for different periods of time, for example, one to six months, six to twelve months and so on. It then worked out the proportion who benefitted over these different periods. The results are given in the following table which shows that the effectiveness improved with the duration of consumption. Some ninety per cent of those who had eaten feverfew for more than two years felt that they were benefitting from it. Thus there is reason to believe that the benificial effects of the plant persist for as long as it is taken.

Proportion of migraine sufferers claiming significant improvement in frequency, severity or duration of symptoms while taking feverfew for different periods

Duration of daily feverfew consumption*	% Improved	% Not improved	Number of users
1 to 6 months	50.0	50.0	66
6 to 12 months	60.6	39.4	33
1 to 2 years	84.2	15.8	57
2 to 5 years	90.8	9.2	66
Overall	72.9	27.1	222

* Those taking feverfew for longer than five years make too small a group for valid comparison.

Is the treatment period indefinite?

Quite a number of migraine sufferers have now taken feverfew for more than five years, and a few that I have been monitoring have taken the plant for more than eight years. So for how long is it advisable to continue treatment? There is no easy answer.

Migraine is known to lessen in frequency and severity in many women as they pass through the menopause. Therefore, women whose periods stop while they are taking feverfew should make sure they really need it before continuing on the plant. They should gradually tail off the leaves over two to four weeks so that as the effects wear off, their body is able to adjust to the plant's chemicals no longer being present. If the underlying migraine process is still active, the attacks will return with their former intensity.

Men and women who have taken feverfew for more than three years would, in any case, be well-advised to stop taking the leaves for at least one month each year to check whether it is still having a beneficial effect, as there is absolutely no point in taking any 'foreign' chemicals whether in the form of synthetic drugs or plant drugs, if they are no longer necessary.

Also, the effects of feverfew on the unborn child are unknown and it would be a sensible precaution for any woman who becomes pregnant to stop taking the leaves straightaway.

How does feverfew work?

Although the precise causes of migraine are still unknown, certain theories have been proposed. As was described in Chapter 1, there is a great deal of evidence to suggest that a disturbance in the width of blood vessels is involved in the migraine process. These changes in blood vessel width have been associated with the release in the body of certain chemicals, such as the prostaglandins, noradrenaline, serotonin, and histamine.

My former students Julia Joce, Nisha Kumar and I have discovered in laboratory tests that the effects of these substances are prevented by feverfew, which behaves like the 'anti-drugs' (see page 7) to oppose the actions of the prostaglandins etc. This is particularly interesting as the prostaglandins have also been implicated in the cause of certain forms of arthritis and some arthritic patients have also reported that feverfew relieves their symptoms.

5

Feverfew and Arthritis

Feverfew was used traditionally for easing inflammation, especially that associated with a fever. According to John Gerarde, writing in 1597, it was beneficial for 'all inflammation and hot swellings'. From that time, and perhaps even before, it has been used to alleviate the symptoms of those conditions known collectively as the rheumatic diseases.

The rheumatic diseases

These ailments affect joints, muscles and sinews, and usually give rise to considerable long-term discomfort, pain and stiffness which make the sufferer less mobile. As a few rheumatic diseases are potentially life-threatening, no patient with arthritic symptoms should consider self-medication until a firm diagnosis has been made by a doctor. Family doctors are used to treating rheumatic disorders – which, it has been calculated, account for about a fifth of their work-load.

Rheumatoid arthritis

This is the commonest form. It is characterized by inflammation of joints (arthritis), but also attacks the internal organs.

The cause is not known but, like migraine, rheumatoid arthritis runs in families and women are affected more commonly than men – the ratio is three to one. In the British Isles it affects about one per cent of the population (whereas migraine affects about ten per cent), and it can occur at any age, even in young children.

The condition starts gradually with a painful swelling of one or several of the fluid-containing joints in the wrists, hands, fingers, elbows, shoulders, hips, knees, ankles and feet. Stiffness is particularly noticeable in the early morning when getting out of bed. The swollen joints are often warm and sometimes bulge with fluid containing large numbers of white blood cells and increased amounts of chemicals such as the prostaglandins, which also feature in the migraine process (see page 7).

In addition to having painful, swollen joints the patient feels generally unwell, weak, often depressed and off her food, may become anaemic and possibly develop hard nodules around the elbows, fingers, ankles, the base of the spine, and in the internal organs. In time the hands may become disfigured, often with bent, claw-like fingers or some displacement outwards from the wrist.

Blood tests are made to confirm the diagnosis as a special 'rheumatoid factor' is present in the blood of many patients with established rheumatoid disease; X-ray examination is also made, to follow the development of, or recovery from, the disease.

The symptoms of rheumatoid arthritis can abate spontaneously and a patient who has suffered severe pain even for years may experience a pain-free period for no apparent reason. These are the periods sufferers pray for when they do not respond adequately to available medicines.

Treatment of rheumatoid arthritis

Like migraine, rheumatoid arthritis cannot be rooted out and treatment is directed towards reducing the symptoms with drugs, physiotherapy, splints or surgery. Bed-rest is often of benefit.

Two groups of drugs are used to treat rheumatoid arthritis and these are called respectively the anti-inflammatory and anti-rheumatoid drugs.

Anti-inflammatory drugs

These include aspirin in large doses: indomethacin (Indocid), phenylbutazone (Butazolidine), ibuprofen (Brufen), naproxen (Naprosyn), ketoprofen (Orudis), flurbiprofen (Froben) and piroxicam (Feldene). Most of the anti-inflammatory drugs are believed to work by interfering with the formation of prostaglandins. In general they counter inflammation, lower a raised body temperature back to normal and give pain relief.

Side-effects No medicine is entirely free of side-effects (see Chapter 1) and anti-inflammatory drugs are well-known to have undesirable properties. It is important to mention them, not to alarm or concern the reader but rather to set in context the side-effects of feverfew discussed in Chapter 6. Indeed it will become apparent that feverfew shares some of the side-effects of the anti-arthritic drugs. Assessing the frequency of side-effects occurring is important to the development of any new medicine – drug regulatory authorities will want to know how it compares in this way with existing products.

All of these drugs can cause indigestion due to irritation of the stomach, with occasional blood loss. They may also cause skin rashes and painful mouth ulcers. They have to be used with particular care if the patient is known to have an ulcer in the stomach or small intestine, or is on anticoagulant tablets to prevent blood clotting, otherwise serious haemorrhage might occur.

Side-effects are common with the major anti-inflammatory drugs, indomethacin and phenylbutazone. Indomethacin, for example, frequently causes headache which may be persistent, dizziness and light-headedness. Rarer side-effects include drowsiness, mental confusion, depression, fainting, high blood pressure, difficulty in breathing, increased blood sugar and disorders of the eyes, bone marrow and nervous system.

Side-effects are even more likely if the patient already suffers from allergic disorders such as asthma, liver or kidney disease, epilepsy, parkinsonism or psychiatric disturbances. For these reasons a careful watch is kept on patients receiving long-term treatment with the drug and periodic examinations of the eyes and blood are made.

Phenylbutazone's side-effects are seen in up to forty per cent of patients on the drug. In addition to experiencing nausea, gastric discomfort, ulceration and bleeding, some patients retain fluid in the tissues or suffer high blood pressure, rashes, thyroid enlargement, blood disorders or hepatitis (see page 70).

Indomethacin and phenylbutazone are therefore normally prescribed only if less potent anti-inflammatory pain-killers such as naproxen or ibuprofen have proved ineffective. These latter drugs are generally better tolerated than aspirin. The side-effects of naproxen include gastric discomfort, bleeding, nausea, allergic reactions, rashes, headache and dizziness, but these are relatively mild and infrequent.

Anti-rheumatic drugs

Sometimes, if the aspirin-like drugs do not adequately suppress the painful inflammation of rheumatoid arthritis or if they cause untoward side-effects, the 'second-line' drug-treatments have to be tried. All of these may cause serious adverse effects. They are specifically intended for the treatment of rheumatoid arthritis and related conditions and cannot be used for other types of arthritis. Also, they achieve a full beneficial effect only after four to six months of treatment, a time-course similar to that of feverfew in migraine prevention.

Gold Repeated injections of the soluble gold salt, sodium aurothiomalate (Myocrisin) may cause a marked remission of the symptoms of rheumatoid arthritis but mouth ulcers and skin rashes are not uncommon. Gold can also damage the kidney and suppress the formation of blood cells, a condition that can be fatal, so blood and urine tests have to be made at frequent intervals. It is not known how gold exerts its beneficial effects.

Penicillamine (Distamine) This drug is as effective as gold in the treatment of the symptoms of rheumatoid arthritis and has the advantage of being effective when taken by mouth. However, it may cause similar serious side-effects, such as damage of the kidneys and bone marrow, and skin rashes. Penicillamine also causes nausea, loss of appetite and taste sensation, mouth ulcers and weakness of the muscles.

Other drugs

Other drugs that sometimes have to be used in rheumatoid arthritis include the corticosteroids. These are hormones which may cause severe side-effects such as the thinning of bones and skin, stomach ulcers and stunting of the growth of children. Chloroquine (Avloclor), a drug first used for malaria, but also prescribed for rheumatoid arthritis, can damage the retina of the eye.

If the disease causes arteries to become inflamed, drugs which suppress the immune response of the body may be used such as cyclophosphamide (Endoxana), azathioprine (Imuran) and chlorambucil (Leukeran). Unfortunately they can also prevent the bone marrow from producing blood.

Surgical treatment

If the inflammation of the joints is resistant to treatment, the membrane lining the joint (the synovial membrane), is sometimes removed. This can be done surgically or, if the patient is elderly, by injection into the joint of radioactive Yttrium which destroys the inflamed cells.

If eventually the joints become completely destroyed they are sometimes immobilized by pinning adjacent bones together, or the whole joint may be replaced.

Osteoarthritis (spondylosis)

Osteoarthritis is a common joint condition whose incidence markedly increases with age. Although common in the elderly it can occur in younger age groups as a result of accidents affecting joints. Pain and joint stiffness in one or several joints are the main symptoms but the inflammatory reaction so common in rheumatoid arthritis, is not usually seen. The condition is called osteoarthrosis to indicate the absence of inflammation. It seems to be the result of degeneration of cartilage (gristle) such as the discs separating bones in the knee joint or spine, often due to previous strain or damage. Some varieties of osteoarthritis tend to run in families. The joints in the fingers, which may become knobbly, are often affected as are weight-bearing joints such as in the hip and lower limb. Movement is often restricted and joints may 'creak' on movement.

Osteoarthritis in the neck is called cervical spondylosis and is manifested as pain in the neck on movement. Spikes or spurs of bone may grow on the affected bones and these sometimes press on the nerves as they leave the spinal cord, causing tingling and weakness in the arms.

The discs in the lower back may be damaged, for example, by lifting heavy weights, and the soft centre bulges outwards, often pressing on nerve roots and causing severe pain especially on movement ('slipped disc'). The pain may be felt down the back of the thigh and leg, when it is called sciatica. Chronic back pain caused by a slipped disc is very common. The damage shows up on X-ray as a narrowing of the space between two adjacent back-bones.

Treatment of osteoarthritis

There is no known cure of osteoarthritis and the condition tends to get progressively worse. Cervical spondylosis can be helped by the wearing of a collar. Lower back disc trouble usually responds to bed rest. Occasionally discs have to be removed surgically if the function of nerves is threatened. The hip and knee joints may also be replaced with artificial metal components if the condition is very severe.

Pain of osteoarthritis tends to come and go, and usually responds to the anti-inflammatory drugs such as aspirin, indomethacin, naproxen etc. used for the treatment of rheumatoid arthritis.

Other forms of arthritis

Ankylosing spondylitis is a condition mainly affecting young men. It occurs with only one-tenth of the frequency of rheumatoid arthritis with which it has some features in common. It chiefly affects the joints of the spine which become immobile and is a very painful condition, especially at night. It is also treated with the anti-inflammatory drugs. The anti-rheumatoid drugs do not work.

Psoriatic arthritis is joint inflammation associated with the skin condition psoriasis. It may be indistinguishable from rheumatoid arthritis but takes other severe forms, especially in the fingers. The anti-inflammatory drugs usually help but in severe cases an anti-rheumatoid drug, such as gold, may have to be used.

Association with other illnesses

Arthritis is sometimes associated with illnesses such as infective enteritis, diseases of the intestine such as ulcerative colitis, Crohn's disease and Whipple's disease, and infections of the urinary tract.

Infective arthritis occurs when a joint becomes invaded by fungi, viruses or bacteria including those that infect the throat, intestine, lungs and urinary tract from where they get into the blood stream before arriving at the joint. The appropriate antibiotic is used but pus may have to be removed from the joint by means of a syringe.

Gout or gouty arthritis is a form of arthritis caused by crystals of sodium urate being deposited in the joint cavity. This can be caused by certain drugs and disease states, but often there is an impressive history of gout in the family. The first joint of the big toe is often affected but it can occur in other sites as well.

Gout responds to colchicine, a drug obtained from the crocus (*Colchicum autumnale*), probenecid (Benemid), sulphinpyrazone (Anturan) and allopurinol (Zyloric). Anti-inflammatory drugs such as indomethacin are also effective.

There are many more varieties of rheumatic conditions and the reader requiring further information is referred to the comprehensive review in *Copeman's Textbook of the Rheumatic Diseases*, 5th Edition, edited by J.T. Scott, (Churchill Livingstone, 1978).

I would, however, emphasize once more that before any person considers self-medication of such a disease he or she *must* be sure that it has been properly diagnosed by a medical practitioner to ensure that there is no more serious and undetected underlying condition.

Self help in arthritis

Probably in the region of sixty per cent of arthritis sufferers gain substantial relief from the newer anti-inflammatory drugs which decrease pain and stiffness. Not all respond to the same drugs but a person who does not respond to one may well benefit from another. Unfortunately the symptoms are suppressed only temporarily – a few hours at best.

Aspirin is the oldest drug and still widely used. It does however have the disadvantage of a fairly high incidence of side-effects (see page 11) even with ordinary doses advocated by doctors. Some people become allergic to the drug.

However, some of the newer anti-inflammatory drugs: naproxen (Naprosyn) for example, combine greater effectiveness with a lower incidence of side-effects. They may also be taken less frequently than aspirin. Nevertheless not all patients get full relief from these treatments and may be unable to take some medicines at all if they have additional conditions such as asthma which is often made worse by anti-inflammatory drugs.

It is not surprising therefore, that – as in the case of migraine – sufferers of arthritis have found many ways of helping themselves.

Reducing weight   It is generally acknowledged that being overweight is not healthy. Osteoarthritis tends to affect the weight-bearing joints, those more prone to wear and tear or damage from impact injuries in sporting activities and road traffic accidents. It follows that the greater the weight these joints have to support, the more likely will be the onset of degenerative changes characteristic of osteoarthritis. The inflamed joints of rheumatoid arthritis are similarly aggravated by excessive weight.

Unfortunately, putting on weight is only too easy in arthritic subjects as painful joints tend to make individuals less mobile and less able to take adequate exercise. A sensible diet with a view to substantially reducing weight, especially in cases where the hips or knees are affected, is probably the most important form of treatment in obese arthritics.

Bed rest   As pain and muscular spasm are largely determined by movement and weight-bearing, rest in bed is extremely helpful – especially for patients whose whole constitution is upset by fever or anaemia. However, patients must be encouraged to adopt a good posture and suitable non-weight-bearing muscular exercises must be undertaken.

Posture   Every effort should be made to correct faulty posture. This can be helped by the wearing of a collar or other spinal support.

Heat   The local application of a warm compress or hot water bottle to muscles before exercises reduces stiffness. Sufferers of osteoarthritis who have access to a heated swimming pool find that gentle passive movement of the affected joint under water relaxes the spasm of the surrounding muscles, reduces the pain and may increase the range of movement.

Stress Many tense and anxious individuals complain of pains, particularly in the neck and back. Tension headache may begin with tension in muscles over the spinous processes at the back of the neck. Identification of the cause of the underlying anxiety may alleviate these physical manifestations.

Worry is known to precipitate relapses of rheumatoid arthritis. How this happens is not known but stress can alter the body's chemistry possibly making the joints and tissues more vulnerable to the causative factors. It is easy to appreciate that the tenser the muscles surrounding an inflamed joint, the more painful that joint will be.

Diet The influence of diet on rheumatic disorders has been suspected, even claimed, to be of importance but no good evidence has been proffered in favour of one or other specific diet.

Although diet is no longer considered to be of fundamental importance as a causative factor in gout, some patients do claim that certain wines and foods, especially those containing the uric acid precursors (offal, fish roe, meat extracts etc.) may precipitate attacks. Such patients tend to exclude these substances from their diet.

Physical Aids Arthritic patients spend a good deal of money on physical aids for the home such as electric knives and can openers, pots with modified handles, liquidizers, bath seats, lavatory aids, long handled combs etc. A recent survey at the Centre for Rheumatic Diseases, Glasgow, revealed that these aids for the home provided far greater benefit than prescribed medicine.

Natural Remedies The poor benefit gained by some arthritis sufferers from prescribed medications has led to them trying all manner of 'natural' products, such as an extract of the New Zealand Green-lipped Mussel, *Perna caniculus*, available over the counter at health food stores. It is claimed to be effective in the relief of symptoms of both rheumatoid arthritis and osteoarthritis, and is taken daily in capsules, tablets or granules. However the extract was negative when tested in animal experiments. The results of formal clinical trials on sufferers are awaited with interest.

64

FEVERFEW AND ARTHRITIS

The herbal remedies used by arthritis sufferers all require proper evaluation but the following plants have long been used in the treatment of arthritis:

White Willow   The bark of the white willow contains salicin, a precursor of aspirin. Salicin was isolated from willow bark around 1830. It was used widely in the treatment of rheumatoid arthritis and to lower the body temperature in fever. Bark is gathered from young branches and air-dried. It is then shredded and usually taken as an infusion (tea).

Vernal Grass   Baths and compresses prepared from sweet vernal grass (hay grass) are widely used for rheumatic conditions in continental Europe.

Black Mustard   Compresses made from freshly-ground seeds of black mustard in hot water have been used for muscular pains, but extreme caution is required as the mustard mixture is highly irritant to the skin and can cause a serious inflammatory reaction.

Lavender   A liniment (rub) made from dried lavender flowers has been applied to the painful areas. Rosemary has been similarly used.

St John's Wort   A liniment has also been made from the oil obtained from this plant.

Devil's Claw (Grapple plant)   An infusion made from the tuberous root of devil's claw has long been used in rheumatoid arthritis.

Juniper   An infusion made from the crushed berries of the common juniper used to be advocated daily for one to two months during both spring and autumn for patients suffering from chronic rheumatism. Sometimes the juniper was replaced by shredded dandelion root or dried leaves of the stinging nettle.

With the exception of the active principle of the willow bark and the subjective relief afforded by liniments, herbal treatments have undergone comparatively little evaluation. Indeed, from the information available, their efficacy appears to be poor.

Copper bracelets and acupuncture The wearing of copper bracelets which are cheap to buy, and acupuncture treatment which is expensive to obtain, have been found to be of little benefit.

Feverfew for painful joints

Following the publicity given to Mrs Jenkins' story that feverfew helped her symptoms of osteoarthritis, many users wrote in confirmation. A retired farmer wrote of his wife and himself:

Our complaint is arthritis in the fingers, arms and shoulders. A friend of ours was taking feverfew for migraine. She was also crippled with arthritis but, after a few months, she was partly cured of arthritis and migraine. So we gave it a try for our arthritis which was rather bad with both of us, having had it for about ten to twelve years. Since taking feverfew the pain got less until we had no pain. These last two years have been without any pain from arthritis. In our opinion it is due to taking feverfew. We have never had any side-effects in any way.

A man suffering from cervical spondylosis wrote:

The effects of feverfew on my problem were swift. In a month I was able to stop taking pain drugs and I have not taken any drug since and with leaving out drugs the feverfew effect was even more noticeable. Needless to say the doctors at Grimsby Hospital were astounded.

Following a television programme in which I appeared in March 1983, a man from Stoke-on-Trent sent me this interesting story:

A friend of the family had an accident some years ago causing him to have a steel plate inserted in his head. Ever since he has been subject to terrible head pains, particularly in the morning on getting out of bed.

A week ago I gave him some feverfew plants from my garden. He called at home today to tell me that since the first day of taking the leaves he has had no sign of a headache. He just can't believe it yet, it's so long since he went a day without terrific pains in his head.

Sufferers of more conventional bone and joint pain wrote of their experiments with the plant. Here are just two:

Watchet, Somerset
However, I do have arthritis and have eaten a leaf or two of feverfew daily fairly regularly for several years (along with a date to disguise the taste). This could have helped, as there is no pain anywhere and the stiffness, etc. is very much under control.

and

Bromyard, Herefordshire.
In November 1982 I started taking three leaves daily for arthritis in my right hand, pain extending to my shoulder and some pain in my left hand . . . By February the improvement was such that I discontinued treatment. In about three weeks I felt twinges and went back to my daily dose of three leaves. My arthritis appears to have gone . . .

However, as in the case of migraine, the story wasn't all one-sided. Some individuals gained no benefit and some reported side-effects as before:

Norwich, Norfolk
. . . my aim in trying to get it (feverfew) was to see if it would be beneficial in the treatment of osteoarthritis from which I suffer.

Eventually I was able to obtain the plant locally – but, alas, I still have bad arthritis.

and

Oxfordshire
I should think that it might be very effective, but it proved something of a poison to me. I chewed two small leaves of it and was surprised to find my lips badly swollen quite soon afterwards. I thought that this might be due to some other

cause so, a few days later I ate two more small leaves. My lips again became swollen and cracked to such an extent that a person calling on me thought that I had suffered a bad fall.

Of the arthritis sufferers with whom I have made contact, more than sixty per cent of those who have taken feverfew considered that the plant had reduced their symptoms. The remainder either found it of no help or encountered side-effects which they could not tolerate.

As far as I know, no study of the use of feverfew in the treatment of these conditions has been published and personal testimonies such as those just quoted are of limited value. However, there are more definite indications that the leaves are having some effect in rheumatoid arthritis. I have personally examined a few patients with the characteristic structural changes in the hands referred to earlier (see page 57). One claimed that she had been symptom-free for the five years she had been taking the leaves. Another found her symptoms to recur only when she stopped taking the leaves.

In a pilot study in which I was monitoring the progress of migraine, I switched patients (unknown to them) from capsules containing dried feverfew powder to those containing placebo. Two female subjects experienced severe aches and pains in muscles and joints. In one, the pains remained localized in the muscles of the lower limb and the knee joint, but the other, aged thirty-nine, experienced severe pains in several joints and muscle groups. The pains moved from day to day around her body and were accompanied by marked stiffness of the neck and shoulder joints so that she was unable to turn her head. She also experienced brief but noticeable weakness of her hand grip. In both women the symptoms on withdrawal from feverfew persisted for almost two months before disappearing.

When 164 migraine-suffering feverfew users who had stopped taking the herb were asked whether they had suffered such symptoms, about one-tenth said that they had. It may be coincidental but this is exactly the proportion of migraine patients who might be expected to develop rheumatoid arthritis, assuming a causal relationship between the two conditions – migraine affects about ten per cent of the population and rheumatoid arthritis about one per cent.

Experimental evidence supporting the anti-inflammatory activity of feverfew

Feverfew contains chemicals called the sesquiterpene lactones. Related sesquiterpene lactones from other plants have inhibited the development of arthritis in animals and some are more potent than the most active anti-inflammatory drug, indomethacin. And there is no reason to believe that the sesquiterpene lactone compounds isolated from feverfew by Drs Deborah Jessup and Peter Hylands at Chelsea College will not act in a similar way.

Two further observations support the suggestion that feverfew might be beneficial in treating inflammatory conditions. The first, made in my laboratory at King's College, London, was that the sesquiterpene lactones in feverfew keep in check the biological actions of prostaglandins and histamine – substances released in large amounts during the inflammatory process. The second observation, made by scientists in Washington showed that extracts of feverfew actually suppressed the formation of prostaglandins.

Is there a link between migraine and arthritis?

There is no obvious link between migraine and arthritis. Sometimes arthritic changes in the neck bones cause headache, but this is usually a tension or muscle contraction headache, not migraine. However, the involvement of prostaglandins in both migraine and arthritis has long been suspected. Aspirin, which prevents prostaglandin build-up, is the principal drug used to treat both conditions, and in similar amounts. Generally speaking, large doses are required for both conditions, about twelve tablets per day if the symptoms persist. Aspirin-like anti-inflammatory drugs such as naproxen are also being used increasingly to prevent migraine, which is paradoxical as one of their side-effects in arthritic patients is headache. Is there a link here too?

Perhaps the most intriguing parallel of all is the development of painful mouth ulcers in a minority of feverfew users and, more commonly, in those treated with both anti-inflammatory and anti-rheumatoid drugs. There the similarity ends however, as mouth ulcers are a minor side-effect of the anti-rheumatoid drug yet the most serious effect reported with feverfew.

This brings us to the question of the safety of feverfew. Like the majority of plants, feverfew has not been tested for its ability to produce toxic effects in animals. Yet a substantial amount of work has been done on humans who use it.

NOTE:
In March 1984, because of the serious side-effects associated with phenylbutazone, the UK DHSS restricted its use to the treatment of ankylosing spondylitis and limited its supply to hospitals only.

6

Is Feverfew Safe?

The serious birth defects that resulted when pregnant women took the sedative drug Thalidomide twenty years ago still haunt the drug industry. It produced one of the most grotesque congenital malformations known, phocomelia or 'seal limbs', in which the bones of limbs are absent or so deformed that the hands or feet appear like flippers on the torso. This first came to light in West Germany in 1961, five years after the introduction of the drug. Eventually the real extent of the tragedy was counted in terms of tens of thousands of malformed babies in some fifteen countries.

Thalidomide had been prescribed widely, and even sold over the counter without prescription, for a relatively trivial condition and a horrified public rightly demanded to know how this had come about. Why was the drug not tested on pregnant animals before marketing? Why hadn't these appalling defects come to light during the early stages of testing in man? Who was to blame? Why had the US Food and Drug Administration (FDA) not approved the marketing of Thalidomide when so many other countries had?

In 1962, largely as a result of questions such as these, a joint sub-committee of the English and Scottish Standing Medical Advisory Committees recommended the establishment of an expert committee to review the evidence available for the safety of new drugs. The Committee on Safety of Drugs came into being in 1963 under the chairmanship of Sir Derrick Dunlop, and drug companies voluntarily submitted their evidence to the 'Dunlop Committee' as it came to be known. Eventually the Committee was given statutory backing and the *Medicines Act* received the Royal Assent in October 1968.

The Act controls all aspects of medicines for human and veterinary use, including marketing, manufacture and distribution. It also makes provision for the establishment of expert committees. One of these, The Committee on Safety of Medicines (CSM), continues the work of the former Committee on Safety of Drugs.

71

The CSM consists of experts, mostly academics, who are independent of the drug industry and its main work is to assess the safety, quality and efficacy of medicines intended for human consumption. No drug can be marketed without first gaining the approval of the CSM. But if this body makes a wrong decision – and harmful drugs such as Eraldin and Opren *have* passed through its net in recent years – the Act specifically protects it from litigation.

The CSM may well have prevented some hazardous drugs from being marketed, but individual members would be among the first to agree that the 'safety' experiments that it stipulates must be carried out on animals may not reveal all of the toxic effects that can occur in man. As has been seen recently, harmful drugs cannot be screened out unless they have obvious effects on animals or on the comparatively small number of patients on whom new drugs are tested prior to marketing. The CSM acknowledges the inadequacy of its controls before marketing and has set up a system for the reporting of adverse reactions to newly-released medicines and other post-marketing surveillance.

Paradoxically, while the pharmaceutical industry is obliged to carry out more and more experiments on animals to comply with the CSM's safety requirements, the public is demanding tighter controls on animal experimentation.

What then are the stages in the development of a new medicine and how far does the work done so far on feverfew comply with the requirements of a drug regulatory authority such as the CSM in the United Kingdom or the FDA in the USA?

Tests on animals

The pharmacologist's job is to predict efficacy and safety, on the basis of his findings in animals. A major drug company may produce thousands of new compounds each year. The pharmacologist has about 100 ways of testing these on each of ten species. It would be physically, if not economically, impossible to use all the tests available so he selects a number which, based on his past experience, are most likely to yield useful information. These form the screening programme, and must show up active compounds and reject inactive or harmful ones as quickly as possible.

The first test is commonly made in mice. Groups of mice are given different doses of a prospective new drug and carefully observed to detect any slight changes in movement, pupil diameter, ear colour, rate of breathing, ability to stand up when placed on one side, ability to cling to a rotating rod, and many other departures from the normal which might suggest that the compound has, for example, sedative, antidepressant, muscle relaxant, pain-killing or antihypertensive properties. The results are compared with those obtained with standard drugs.

The drug then undergoes more specialized evaluation of the activity observed in the tests on mice. Complicated tests are carried out on higher mammals and on isolated tissues and at this stage it is decided whether the biological activity discovered warrants testing for safety (toxicity).

Toxicity tests – tests of potential damage due to drugs or chemicals – are of two main kinds: acute tests last for only a short time, chronic tests require the repeated dosing of animals over weeks, months or years. The drugs are administered by the route envisaged for man.

The LD50 (fatality Test)

Acute toxicity tests are made to determine the LD50 value in rats, mice and other species after a single dose and after multiple doses given, say, daily for two weeks. The LD50 is the dose (D) of drug that will kill (L = lethal) fifty per cent of the experimental animals. This dose is then compared with the ED50, the dose of drug (D) that is effective (E) in producing a desired response in fifty per cent of the animals. The LD50 values indicate whether it is worth continuing with the drug. They also indicate the highest dose that can be used in the long-term toxicity tests, and show whether the drug is equally harmful to different species. The number of animals used in the LD50 test can be thirty to sixty.

Chronic toxicity

These long-term toxicity tests of drugs intended to be given by mouth for more than thirty days usually last for twenty-six weeks in the United Kingdom and twelve weeks, followed by a second study of one year, in the USA. Four groups are used of each of two species, usually rats and either dogs or primates. Each group must contain equal numbers of males and females. One group is given the maximum tolerated dose chosen to induce effects on the most sensitive organs and tissues during the period of the study but low enough not to kill most of the animals. The second group is given two or three times the maximum therapeutic dose intended for man. The third group receives a dose in between these two and the fourth group is given no drug but serves as a control for the test groups. In every other respect, the control and test animals are treated identically.

The animals are observed carefully throughout the period of the chronic toxicity testing. The amount of food eaten is recorded, as is body weight. Blood tests of liver, bone marrow and kidney function are made at frequent intervals. After twenty-six weeks (in the UK) the organs of all the animals are examined for dose-related toxic changes.

Teratology (Study of abnormal fetuses)

The drug's potential for inducing fetal abnormalities is looked for in the teratology studies, required by the CSM since the experience with Thalidomide. Two species giving large litters are usually tested – rabbits, rats or mice.

The drug is given throughout pregnancy, commencing immediately after mating. Again, four groups are used, the control group being used to assess the *natural* frequency of malformations occurring. At the end of pregnancy the fetuses are examined for any deformities or abnormalities.

Fertility

Another statutory test is that of fertility and general reproduction in which both males and females are dosed prior to mating. Rodents are preferred for this test and again, three dose groups and one control group are included. These tests are designed to discover whether a drug suppresses fertility or induces changes in the development of spermatozoa. Many of the animals are allowed to deliver and some offspring are allowed to mature, then mated so that a second generation can be studied for changes in fertility.

Mutagenicity (Ability to alter chromosomes)

The mutagenicity test shows whether a drug is able to induce changes in the structures called chromosomes, through which the genetic code is transmitted from parents to offspring or – in the case of multiplying bodily cells – from parent cell to daughter cells. The test can be made on bacteria, mammalian cells kept alive in culture or in animals. The bacterial test is the most usual. Positive results sometimes indicate that the drug might cause cancer.

Carcinogenicity (Ability to induce cancer)

Tests of the ability of drugs to induce cancer are known as carcinogenicity tests. These are mandatory if patients are likely to receive the new drug for more than one year. In any species tumours take a long time to develop and may arise spontaneously. For these reasons large groups of test animals have to be used (fifty per cent of each sex per group) and a large control group is included (100 of each sex). Animals taken just after weaning are dosed daily for the rest of their lives – eighteen months for mice; two years for rats; eight years for dogs and so on. The top dose given is the highest that will permit growth: the other doses are as for the six months chronic toxicity test.

Two species have to be used which handle the drug in a manner similar to man, where this is known. Furthermore, they must be sensitive to known cancer-inducing chemicals. The UK Department of Health and Social Security stipulates the microscopical examinations that must be made at the end of this study, and the drug companies have to retain specimens for many years after the tests have been completed.

The first clinical testing on humans can be initiated before the results of the carcinogenicity studies are known. But first, all of the data from the animal studies has to be assembled along with data on purity and stability of the drug chemical and how it is formulated into tablets or capsules. This information is sent to the CSM with a request to undertake clinical trials.

Tests on human volunteers

The clinical investigation of a new drug is generally subdivided into four stages.

Stage 1 is designed to confirm in man the safety results obtained in animals. In this phase the drug is given to man for the first time. As the first dose given is usually less than that considered likely to benefit a patient with a disease, it is normally given to a healthy volunteer. Not only are we interested in whether the drug induces any anticipated beneficial change, such as a lowering of blood pressure or slowing of the heart beat but, more importantly, tests are made to see if there are adverse effects on the body. This latter is determined by the examination of blood and urine samples for abnormal constituents. These tests are likely to pick up early cellular and chemical signs of liver, kidney and bone marrow damage before symptoms become apparent. If the tests are negative, the next volunteer receives twice the initial dose and the whole procedure is repeated.

When, eventually, the doses given are within the range envisaged for treatment, a volunteer may be given more than one dose on successive days. Stage 1 ends with the repeated dosage of actual patients, using the therapeutic dose and regimen calculated from the studies carried out on healthy volunteers. All the safety data gathered from the blood tests are analysed before proceeding further. The number of subjects (both healthy volunteers and patients) in Stage 1 will usually be in the range of twenty to fifty who receive the new drug.

Stage 2 is when substantial clinical trials are carried out in patients. These are designed to demonstrate therapeutic efficacy and to confirm the safety results of Stage 1. In this phase the therapeutic dose is established with a fair degree of certainty, as is the dose that produces toxic effects. Stage 2 seldom includes more than 200 patients. These are closely monitored and full blood tests are made from time to time. The trials are also usually controlled – that is, the new drug is compared with either an established treatment of a placebo (see page 10).

Stage 3 More patients receive the drug in controlled and uncontrolled trials. The evidence of efficacy gained in Stage 2 is enhanced and further evidence of safety, tolerance and a more precise definition of side-effects is obtained. Up to 3000 patients may have received the drug by the end of this Stage. In the UK, the CSM usually requires at least 100 patients to have taken the drug continuously for more than one year if the treatment is for a condition such as migraine or high blood pressure, for which it will be used continuously over several years.

At the end of Stage 3 all the clinical results are submitted to the Licencing Authority with an application for a Product Licence. Once this is granted, the drug can be marketed.

Stage 4 trials take place after marketing has commenced. They include investigation of the rarer adverse reactions, especially those which may occur only when the patient has received the drug for a long time. An example of such a delayed adverse effect is the eye damage caused by the drug Eraldin. Often, the new drug is also tested to see if it has specific advantages over well-established competitive products.

Safety of feverfew

To my knowledge, no chronic toxicity tests have yet been made on feverfew because its use as a preventive treatment of migraine was initiated by sufferers who used it and not by the drug industry. Because man was the first experimental animal on which feverfew was tested, its development has been exactly opposite to that of conventional new medicines.

As far as I can tell from the long-term users I have examined and whose blood has been tested, no serious side-effects have yet emerged. I would emphasize *yet* because three to five years is not long in the life of a drug; for example, the effects of thalidomide were not identified until five years after marketing. Furthermore, feverfew is still probably consumed by just a few thousand individuals. Often, the rarer adverse effects only become evident when hundreds of thousands have taken a new medicine.

However, my colleagues and I have carefully documented information on more than 300 feverfew users, representing in excess of 600 patient-years of daily treatment. This is probably more than is documented for many new medicines by the time that application is made for a product licence and certainly more information than is available on most herbal products.

Although the LD50 value (see page 73) has not been esti-mated, rats trained to eat their normal diet in powdered form have been given measured amounts of feverfew powder mixed in. They readily ate every day more than 100 times the human daily dose. Furthermore they did so for five successive weeks without any loss of appetite or difference in weight gain when compared with control animals given the powdered normal diet only.

Similarly, guinea-pigs ate about 150 times the human dose every day and after seven weeks were identical in every respect with control animals – their coats looked as sleek and their weight gain was the same.

Feverfew therefore satisfies that fundamental requirement of all new drugs: that the dose required to produce a therapeutic effect in man is less than one-hundredth and, preferably less than one-hundred-and-fiftieth of the dose that will cause serious toxic effects. Although the toxic dose of feverfew has not yet been determined in animals, judging by the observations de-tailed above, it must be extremely high.

Blood tests in long-term users

Just as many illnesses cause changes in the cellular and chemical composition of blood, so can the toxic actions of drugs and chemicals. Blood tests are therefore of vital importance in patients receiving new drugs.

Some blood chemicals are present in different concentrations in women than men – blood fats for example – and some blood enzymes increase in concentration with age. For these reasons it was not sufficient merely to test the blood of patients taking feverfew, but also to control test for changes due to age, sex and migraine, by examining blood samples from other groups, matched for age and sex but who were not eating the plant. All participants were first examined to exclude other illness and tests were carried out on the following four groups, each of thirty individuals:

1   Migraine sufferers who had taken feverfew daily for more than three months, but preferably for more than one year.

2   Migraine sufferers who used to take feverfew daily but who had stopped taking it at least three months before.
3   Migraine sufferers who had never taken feverfew.
4   Healthy volunteers who had never suffered from migraine or taken feverfew.

The blood of each of the 120 individuals was tested for forty separate constituents. Although a few abnormal values were obtained, the four groups did not differ in any significant respect so no evidence emerged that implicated feverfew in causing damage to any vital organ. Certainly, in my opinion, the amount of safety data obtained from this study more than equalled that expected on completion of Stage 1 of the clinical evaluation of a new drug in man.

Side-effects of feverfew

The assessment of the incidence and severity of side-effects caused by drug treatments is an important part of all clinical studies. Side-effects can be defined as those *unwanted* pharmacological actions of a medicine, occurring with doses normally used to treat a condition. Examples include drowsiness from tranquillizers such as diazepam (Valium), vomiting with the antimigraine drug ergotamine (Migril) and dizziness on some drugs which lower high blood pressure too quickly. Side-effects to feverfew occurred in only 17.9 per cent of 270 patients surveyed. None was serious in the sense that it was life-threatening. They included:

Side effects	Incidence (%)
Mouth ulcers/sore tongue	6.4
Abdominal pain/indigestion	3.9
Unpleasant taste	3.0
Tingling sensation	0.9
Urinary problems	0.9
Headache	0.9
Swollen lips/mouth	0.4
Diarrhoea	0.4

Exactly two-thirds of all those who remembered when their side-effects began, said they were apparent within the first week of treatment. Curiously, when those suffering from migraine only were asked for the same information, two-thirds said their side-effects came on gradually over the first two months and only a quarter experienced them in the first week of treatment.

When feverfew-users were asked whether they had ever experienced specific side-effects such as ulceration of the mouth or indigestion while taking the leaves, a higher proportion said that they had. For example, 11.3 per cent had suffered from mouth ulcers and 6.5 per cent had experienced indigestion in comparison with 6.4 per cent and 3.9 per cent respectively who volunteered these side-effects.

Mouth ulcers

Some people are known to have more sensitive skin than others, and rashes can arise when their skin comes into contact with certain plants such as the primula. Contact dermatitis, as it is called, has also been reported with feverfew. Also, some people get skin rashes because they are allergic to certain foods they eat, such as plums and strawberries. It was of interest to know, therefore, whether the mouth ulcers occurred because a sensitive mucous membrane had come into direct contact with the leaves during chewing or because of a 'systemic' effect following absorption of the plant products.

Users of raw feverfew leaves had always assumed that mouth ulceration was a contact response but this was easily disproved. When I transferred a 39-year-old patient from feverfew leaves to capsules containing dried feverfew she still suffered mouth ulcers. When, unknown to her, placebo capsules were substituted her ulcers cleared up within a week.

Some people who start to take feverfew capsules containing a high dose of dried leaf (100 mg or over), obtained through the post or from certain 'health food' retail outlets, have written to the migraine charities to say that they get dreadful mouth ulcers. So buyers beware! The feverfew capsules available at the time of writing have been marketed without adequate clinical testing to support the dose advocated, or its safety.

Ulceration of the mouth is the most common troublesome side-effect attributed to feverfew and it is most probably due to the actions of one or more substances absorbed from it after it has been eaten. Experiments are now in progress in susceptible individuals to see if a marked reduction of dosage will prevent the ulcers from forming while still controlling the migraine.

Some patients think that because feverfew causes mouth ulcers it most probably causes stomach or intestinal ulcers as well. Certainly, the aspirin-like drugs can do this and one of their more common side-effects is a slight blood loss from the site of ulceration. So the possibility that feverfew might cause blood loss and hence lead to anaemia has certainly been considered when interviewing and examining patients. Happily there is no suggestion so far that feverfew-induced ulceration of the mouth is paralleled by ulceration elsewhere.

Of the three women I discovered to be anaemic on clinical examination and subsequently confirmed by blood tests, one had an hereditary blood condition called Thallassaemia in which the red corpuscles are wrongly formed and break up, the other two had chronic iron deficiency anaemia most probably due to inadequate dietary intake of iron. In neither case was there a history of symptoms suggestive of ulceration of, or bleeding from, the alimentary tract.

Mouth ulcers are caused by other drugs and they are commonly encountered with most of the drugs used in the treatment of rheumatoid arthritis: the anti-inflammatories, gold and penicillamine. As in the case of feverfew, the ulceration is caused by these drugs after they have been absorbed from the intestine or site of injection. Whatever else can be said of feverfew, its ability to induce mouth ulcers in such a remarkably low daily dosage, shows that the plant possesses marked biological activity which needs further detailed investigation.

Feverfew also sometimes induces a more widespread inflammation of the mucous membrane of the mouth and surface of the tongue, often with swelling of the lips. This could be caused by direct contact with the plant during chewing. In its most severe form there can also be loss of taste sensation. This side-effect tends to occur in susceptible individuals after several weeks or months of daily use, as in the case of a woman writing from Croydon:

A couple of years ago I ate two leaves every day and was delighted with the results. My headaches became very infrequent and I happily cultivated feverfew plants so that I was never without a supply.

However, after nine months of relative freedom from headaches I stopped eating the leaves. I had side-effects which for me outweighed the advantages.

I lost all my sense of taste and my tongue developed deep fissures and a very ragged appearance. Possibly this is something peculiar to me . . .

Many of the people now using the plants I have passed on, take their leaves as a drink or in a sandwich so that contact with the tongue is minimal and so far I haven't heard of any side-effects from them . . .

Reasons for stopping feverfew treatment

Virtually every domestic medical cabinet contains up to three prescription medicines that should have been taken as directed but were not, for a variety of reasons. Most people do not like taking medicines for long periods. They are suspicious of 'foreign' chemicals, of possible long-term effects, of lack of benefit, or they may have experienced side-effects not acceptable to them. Thus it seemed advisable to find out why people stopped taking feverfew in order to see what side-effects are serious enough to lead to discontinuation.

At the time of our analysis 164 users had stopped taking feverfew leaves. Their reasons are given in the following table, from which it can be seen that the vast majority of reasons were not directly attributed to the side-effects of the plant. The principal reasons were poor efficacy (thirty per cent) or reasons unrelated to the plant as a medicine, such as the inability to obtain leaves (twenty-four per cent) especially in winter. Anxiety over unfounded suggestions in the press of potential side-effects accounted for ten per cent. Others stopped the leaves because their migraine improved, only to recur. Only twenty-one per cent of those who stopped eating feverfew did so because of side-effects, the majority of which affected the gastro-intestinal tract.

Feverfew and other drugs

So far no clear-cut adverse interactions have emerged between feverfew and drugs used to treat additional medical conditions

IS FEVERFEW SAFE?

Reasons why 164 users stopped taking feverfew*

	Number of Reports	% Subjects
Poor efficacy	49	29.9
Migraine or headache provoked	2	1.2
Migraine improved or stopped	21	12.8
Intercurrent illness/hospitalization/death in family	7	4.3
Experiment to see if migraine returns	3	1.8
No supply	40	24.4
Anxiety over reports of side-effects	16	9.8
To try other treatment	1	0.6
Doctor's suggestion	2	1.2
Different diagnosis made	1	0.6
No reason given	4	2.4
Became post-menopausal and migraine ceased	1	0.6

Side-effects

A *Gastrointestinal*		
Mouth ulcers	7	4.3
Swollen lips/throat/tongue	3	1.8
Dry/sore tongue	2	1.2
Nausea and/or vomiting	4	2.4
Bitter taste	1	0.6
Heartburn/indigestion/stomach upset	5	3.0
Abdominal pain	2	1.2
Diarrhoea	1	0.6
Flatulence	1	0.6
B *Miscellaneous*		
Skin rash	1	0.6
Provocation of menstruation	1	0.6
Allergy	1	0.6
Disturbance of vision	1	0.6
C *Unspecified*	5	3.0
Total Reports	182	

* More than 1 reason was offered by a few individuals.

from which migraine patients were suffering (see Chapter 4). However, it is reasonable to suppose that some of the medicines and pain-killers for example, would enhance the beneficial effects of feverfew in migraine.

The drugs which have been co-administered with feverfew, excluding those used for migraine, are listed in the table below. The absence of obvious unfavourable interactions with such a large number of diverse drugs is encouraging. Naturally, a watch for potential interactions between feverfew and other drugs must be continued for a long time yet. No adverse interactions have yet been recorded between feverfew and almost 50 different drugs and drug combinations used for the acute treatment or prevention of migraine. Nor does there seem to be any suggestion that feverfew is more likely to cause mouth ulcers or other side-effects when taken with other medicines.

Drugs used simultaneously with feverfew but taken for other medical indications

	Number of users
Analgesics	
Dextromoramide (Palfium)	1
Dextropropoxyphene + paracetamol (Distalgesic)	2
Paracetamol (Panadol)	2
Paracetamol + codeine (Paracodol)	1
Antacids and anti-ulcer drugs	
Alginic acid + magnesium trisilicate + aluminium hydroxide + sodium bicarbonate (Gaviscon)	1
Ambutonium + aluminium hydroxide (Aludrox)	1
Carbenoxolone + magnesium trisilicate + aluminium hydroxide + sodium bicarbonate (Pyrogastrone)	1
Cimetidine (Tagamet)	4
Dimethicone + aluminium hydroxide (Asilone)	2
Hydrotalcite + activiated dimethicon (Altacite)	2
Oxethazaine + aluminium hydroxide + magnesium hydroxide (Mucaine)	1
Poldine methylsulphate (Nacton)	2
Anti-anxiety drugs and sedatives	
Clorazepate (Tranxene)	1
Diazepam (Valium; Sedapam)	4
Dichloralphenazone (Welldorm)	1
Flurazepam (Dalmane)	1
Lorazepam (Ativan)	2
Nitrazepam (Mogadon)	3

Anti-anxiety drugs and sedatives (cont.)	Number of users
Oxazepam (Serenid D)	1
Triazolam (Halcion)	1
Trifluoperazine (Stelazine)	1

Anticoagulants

Warfarin	1

Antidepressants

Amitriptyline (Tryptizol)	2
Dothiepin (Prothiaden)	1
Imipramine (Tofranil)	2
Tranylcypromine + trifluoperazine (Parstelin)	1

Anti-emetics

Cinnarizine (Stugeron)	1
Metoclopramide (Maxolon)	1
Prochlorperazine (Stemetil)	2

Antihistamines and anti-allergy drugs

Chlorpheniramine (Piriton)	1
Cromoglycate (Intal)	2
Diphenhydramine + ammonium chloride + menthol + sodium citrate (Benylin)	1
Grass pollen extracts (Pollinex)	1
Ketotifen (Zaditen)	1
Methylprednisolone (Medrone)	1
Prednisolone (Prednesol)	2
Prednisone (Delta cortone)	1
Triprolidine (Pro-actidil)	2
Triprolidine + pseudoephedrine (Actifed)	2

Antihypertensives

A   *Beta-blockers*

Atenolol (Tenormin)	2
Labetalol (Trandate)	2
Propranolol (Inderal)	7

B   *Diuretics*

Bendrofluozide (Aprinox; Neo-naclex)	2
Cyclopenthiazide + potassium (Navidrex K)	4
Frusemide (Lasix)	1
Spironolactone + hydroflumethiazide (Aldactide)	1
Xipamide (Divexan)	1

C   *Others*

Debrisoquine (Declinax)	1
Methyldopa (Aldomet)	1

FEVERFEW

Antimicrobial drugs	Number of users
Amoxycillin (Amoxil)	1
Antibiotics (unspecified)	3
Cotrimoxazole (Septrin)	1
Ketoconazole (Nizoral)	1
Nystatin (Nystan)	1

Bronchodilators

Aminophylline (Cardophylin)	1
Beclomethasone (Becotide)	2
Salbutamol (Ventolin)	2
Terbutaline (Bricanyl)	1
Theophylline (Nuelin)	1

Drugs acting on the gastrointestinal tract

Codeine phosphate	1
Danthron + Polaxamer 188 (Dorbanex)	1
Diphenoxylate + atropine + neomycin (Lomotil)	2
Hyoscine (Buscopan)	1
Ispaghula husk (Fybogel; Isogel)	3
Mebeverine (Colofac)	2

Haematinics

Ferrous sulphate (Feospan)	1
Folic Acid	1

Hormones

Conjugated oestrogens (Premarin)	5
Norethisterone + ethinyloestradiol (Orlest 21)	1

Miscellaneous

Allopurinol (Zyloric)	1
Carbamazepine (Tegretol)	1
Flavoxate (Urispas)	1
'Homoeopathic medicine' (unspecified)	1
Phenobarbitone (Luminal)	1

Non-steroidal anti-inflammatory agents

Diclofenac (Voltarol)	3
Ibuprofen (Brufen)	3
Naproxen (Naprosyn)	1
Phenylbutazone (Butacote)	1
Piroxicam (Feldene)	1

7

The Experimenters

A medicine seldom possesses only one pharmacological property which can be used in the treatment of a single disease. Most have several actions, thus enabling them to be used in the treatment of a number of conditions. Aspirin for example, is used not only in treating rheumatoid arthritis and migraine but also to lower the body temperature in fever and sometimes to thin the blood in those prone to heart attacks. The drug propranolol is used primarily in the treatment of high blood pressure, but also in the treatment of some heart conditions characterized by an irregular pulse, the painful heart condition known as angina pectoris, certain forms of tremor, control of anxiety and prevention of migraine. Oral contraceptive pills are used not only to prevent pregnancy but sometimes to control premenstrual tension, painful periods and heavy bleeding from the womb. Herbal medicines, too, can have more than one pharmacological action and can occasionally be used to treat a number of conditions.

Additional beneficial effects of feverfew

Migraine patients taking feverfew attribute a number of pleasant effects to the plant, apart from the reduction in frequency and intensity of headaches. These include the prevention of nausea and vomiting, relief from symptoms of arthritis, a sense of well-being, more restful sleep and improved digestion. In all, fifty-three per cent of feverfew users claimed that the plant's consumption gave rise to pleasant side-effects and a minority of people seem to be very sensitive to them.

Whereas the unwanted side-effects of feverfew appeared early, the additional beneficial effects tended to coincide with the antimigraine response in speed of onset. Thus, one in six noted them during the first week, one-third by the end of the first month and two-thirds by the end of the second month. Two-thirds of individuals who encountered adverse side-effects did so within one week of starting to use feverfew. Furthermore, once the beneficial effects occur, they continue for as long as the plant is taken. I have come across only two instances where this was not the case.

Tension and depression

In a double-blind trial, an elderly patient whose migraine had apparently spontaneously disappeared while he was taking feverfew leaves felt extremely nervous and on edge during the first two months on placebo capsules, even though his migraine headaches did not return. He recalled feeling like this before he started taking feverfew. He was quite certain that feverfew had a calming action and that he was, therefore, on placebo. This calming property has been noted by many others, which brings to mind a use of feverfew in the treatment of depression as described by the herbalist Nicholas Culpeper in 1653:

> The powder of the herb taken in wine, with some Oxymel . . . is available for those that are short winded, and are troubled with melancholy and heaviness, or sadness of spirits.

And John Gerarde wrote in 1597:

> Feverfew dried and made into pouder, and two drams of it taken with honie or sweet wine purgeth by siege melancholie and flegme.

Menstruation

Feverfew has been used in the past as an emmenagogue, that is, an agent which assists and promotes menstrual flow.

Again Gerarde:

> . . . it procureth womens sickness with speed, it bringeth forth the afterbirth and the dead childe . . .

and Culpeper:

> It cleanses the womb, expels the afterbirth, and doth a woman all the good she can desire of an herb . . . and (a decoction) drank often in a day, is an approved remedy to bring down

women's courses speedily, and helps to expel the dead birth and after-birth.

This ability to cause abortion seems also to apply in animals – feverfew is known to cause abortion when eaten by cows.

This brings us face to face with yet another coincidence. Ergotamine, the drug used in migraine, will also contract the womb and expel the afterbirth. Ergometrine, a drug chemically related to it, is now used for this specific purpose. Both drugs come from the fungal plant ergot, a parasite of the rye grain, and crude extracts of ergot have been used for centuries in midwifery.

The anti-inflammatory drugs such as indomethacin are now becoming increasingly popular in the treatment of painful periods due to their antiprostaglandin effect. Perhaps herein lies an explanation for feverfew's action on the womb.

However, when feverfew was used to assist menstruation it was given in high doses, and is no longer commonly administered in this way. Even so, I always enquire about changes in menstrual flow in women of child-bearing years and so far it doesn't seem to be an important problem for the majority. In fact I know of only one person who stopped taking feverfew because it made her periods worse, but I did meet a most striking example of how feverfew can affect menstruation in a forty-year-old woman from Oxfordshire who participated in the same study as the man who felt nervous on placebo capsules. She detected, or rather guessed correctly, that she was on placebo because her periods were lighter and less regular on the capsules and she had noticed this change when she came off feverfew once before.

So, if feverfew can modify menstrual flow, cause abortion in cattle and – according to the herbals – induce uterine contraction in full-term women, it would seem prudent for pregnant women not to take it until more is known of its pharmacological actions, especially as little is known of its effects on migraine and arthritis in pregnancy.

Feverfew in the treatment of other conditions

There seems no shortage of sick people who are prepared to experiment with feverfew for a variety of conditions, some of which have apparently responded markedly to the same treatment regimen used by migraine sufferers. Some are recorded here.

Ménière's syndrome

Ménière's syndrome is a condition of unknown cause but characterized by recurrent bouts of severe dizziness or vertigo associated with ringing in the ears (tinnitus) and progressive irreversible deafness. Many sufferers also have a history of migraine. One such individual, a 79-year-old woman from Cardiff, completed the Feverfew questionnaire and gave a detailed description of the response of the condition which she had suffered for many years.

The giddiness was so severe that sometimes the patient was unable to stand. It was accompanied by severe nausea, vomiting, sweating and faintness. She obtained a complete and lasting remission on three small leaves of feverfew taken every day, and this remission has persisted for over two years. The sickness and vertigo have been more or less eradicated; before taking feverfew they used to last for hours. She now takes no other medicine for her Ménière's.

A result as striking as this must have been observed before. So what did Gerarde and Culpepper have to say? First Gerarde:

> . . . it is very good for them that are giddie in the head, or which have the turning called Vertigo.

Now Culpeper:

> It is very effectual for all pains in the head coming of a cold cause . . . : As also for vertigo, that is a running or swimming in the head.

There is no consistently good treatment of Ménière's syndrome, although antihistamines such as chlorpromazine (Largactil) and prochlorperazine (Stemetil) given three times every day sometimes reduce the number of attacks. As extracts of feverfew have also been found to have potent antihistamine activity, there may be a rational basis for its apparent beneficial effect in this patient.

High Blood Pressure

It has been suggested by feverfew users that high blood pressure may respond favourably to the daily consumption of the leaves. Thus a Gloucestershire woman wrote:

> Until two years ago I was having migraine attacks at about fortnightly intervals and spent a great deal of time in a darkened room in considerable pain.
>
> The doctor's tablets caused me to lose the feeling in my fingers.
>
> I read about feverfew – was sceptical about its effectiveness – tried it during an attack with no result. Later I tried it again, this time taking it regularly just after breakfast. Since then I have only had two attacks bad enough to need a day in bed and even then the pain was dull and bearable as opposed to the throbbing I used to get . . .
>
> I grow the feverfew, dry it and grind it up. The powder is easy to take with a cup of tea . . .
>
> I wonder if you can tell me anything about its chemical content. A friend and myself were both told by different doctors that our blood pressure, which did not need treating, though at the limit, had unaccountably lowered.

The loss of sensation in her fingers was presumably caused by ergotamine and is, in fact, a symptom of ergotamine overdose. Ergotamine, by constricting the blood vessels, also tends to cause the blood pressure to rise, so the discontinuation of ergotamine alone could account for the fall in blood pressure. Having said this there are theoretical pharmacological reasons for thinking that feverfew might lower blood pressure. But I have no evidence that this occurs – and the blood pressure of rats fed high doses of feverfew for long periods of time does not change significantly. However, the only certain way to determine the blood pressure lowering effect of feverfew is to measure it in those who are about to start taking the plant and then repeat the measurements at intervals over the following three to six months.

Respiratory Disorders

Feverfew has been used for centuries for the treatment of respiratory illnesses. Thus, according to Culpeper:

> The decoction thereof made, with some sugar, or honey put

thereto, is used by many with good success to help the cough and stuffing of the chest, by colds, . . . The powder of the herb taken in wine, with some Oxymel, purges both choler and phlegm, and is available for those that are short winded . . .

Dodoens, writing in 1578, said:

Feverfew dried and made into pouder, and two drammes of it taken with hony, or other thing, purgeth by siege Melancholy and fleume: where for it is very good . . . for them that are purse or troubled with the shortnes of winde, . . .

Feverfew's traditional use for the treatment of colds is indicated by the following extract of a letter to the British Migraine Association:

Peel, Isle of Man

My great grandmother was a herb woman and knowledge of this plant has been passed down by word of mouth. But our granny grew it in a London garden, and she picked the plant, leaves and flowers, and poured boiling water over it. When cold we drank it in a small wine glass. This was when we had bad colds and headaches with stomach upsets.

The 'headaches with stomach upsets' is most probably a reference to the folk use of feverfew in the treatment of migraine.

In the laboratory feverfew extracts have been shown to prevent the spasm of muscles such as those found in the respiratory tract. Perhaps this explains why some individuals find it of help in asthma ('shortnes of winde'?).

Brynmenyn, Nr. Bridgend

I have suffered from asthma for over 20 years, and believing there is a link between migraine, asthma and eczema, I started taking feverfew . . .

After taking feverfew since 1978 (4 years), the result is remarkable. I have been able to reduce the amount of cortisone I have been taking for many years and feel altogether better. I used to be a wheezing invalid, but no longer, thanks to feverfew.

Others claim benefit from ordinary chrysanthemum leaves as did this 73-year-old gentleman:

Ipswich, Suffolk

I read in the *News of the World* about eating chrysanthemums to relieve headaches. I have a cure for asthma given to me years ago. Fill a two pint jug with chrysanthemum leaves and cover with boiling water, and leave to soak. Strain and drink the liquid.

Insect bites

Mrs Lesley Hirst, whose recipe for feverfew pills appears on page 36, made the following interesting observations. Before she took feverfew she used to react violently to insect bites, especially gnat (mosquito) bites which caused marked inflamed swelling. Since taking feverfew she never seems to react to insect bites and wonders whether she is now no longer being bitten.

I think the likely explanation is that Mrs Hirst is being bitten just as often but that the inflammatory reaction to the insect bites is being suppressed by chemicals absorbed from the feverfew. The antihistamine action of the plant may be particularly important in this respect.

It may also possess insecticidal activity, as members of the *Chrysanthemum* and *Tanacetum* genera are well-known to contain chemicals protective against insect attack. Tansy, in particular, is still kept in the dried form to repel insects. At one time it was even used in the embalming of corpses.

Other conditions

Several other experimenters were prepared to try feverfew for more serious conditions. I would mention four of them only by way of *discouraging* other would-be users with the same disorders.

A Hampshire correspondent took feverfew for brucellosis or undulant fever, a bacterial infection transmitted from cattle to man, via unpasteurized milk. She took feverfew for some time then, fortunately in her case, 'saw an article in *Woman's Weekly* (see page 22) which said it was toxic to the liver', and stopped taking it.

So Doctor Margaret's article did some good after all! Brucellosis, once diagnosed, should be treated only by the appropriate antibiotics, otherwise the symptoms of sweating, weakness, headache, loss of appetite, pain in the limbs, joints and back, cough and sore throat may recur intermittently for many months.

A man also from Hampshire, experimented with feverfew for twelve months for a serious attack of glaucoma with no improvement.

Glaucoma is a condition of the eye characterized by an increase in the pressure of fluid within the eyeball. The pressure can be so high that the retina at the back of the eye can be damaged and blindness can result. As effective treatments are available, no patient with diagnosed glaucoma should experiment with feverfew or any other drug or herbal preparation, without medical supervision.

The third condition I would mention is intermittent claudication – extremely severe pain in the calves on walking due to impairment of the circulation in one or both legs. No treatment seems to be particularly effective in this condition and I wouldn't expect feverfew to help it either unless by preventing some of the associated calf muscle spasm, despite the following letter received from a Norfolk man:

I am 66 years of age, and for a few years have had a problem with my leg, inasmuch as, after sitting, I would get up and walk a short distance and a pain came into my calf, and on stopping for a short time, would disappear but would repeat itself every time I sat down.

I attended a meeting with a medical consultant who informed me that this was due to too many cholesterols in my blood stream, and put me on a low fat diet. It appeared to ease the problem for a time but then it came back as bad as ever.

My wife and I spend our holidays on the Isle of Wight with a very nice couple, and during a conversation, my leg problem arose, when it was suggested that I tried feverfew, three or four leaves each day. Whether this is helping my problem, or not, I must say that my leg appears to be getting a lot better.

Lucky man! But in my view the improvement is unlikely to be due to feverfew.

I have had two separate reports on the use of feverfew by patients suffering from multiple sclerosis but know nothing of the outcome in these cases. It is understandable that individuals with progressively deteriorating conditions for which there is no known cure should grasp at any straw, but self-medication of potentially fatal disorders with feverfew should *always* be with a doctor's knowledge and preferably with his agreement.

8

Today's Unorthodoxy
Tomorrow's Convention?*

The recent renewal of interest in feverfew stems largely from three articles written for doctors. Two were compiled by a young journalist, Julian Chomet, for the medical newspapers *General Practitioner* in May, 1982 and *Current Practice* in February, 1983. The third was a brief paper which I wrote for *Mims Magazine*, in May, 1983 in a special number devoted to migraine. Considerable interest was shown in this article and summaries appeared in local, national and international newspapers.

Like most doctors I feel that the correct place for discussion of medical matters is provided by the medical periodicals. But the public has a legitimate interest in the advances made in medical science and it is reasonable that those associated with such advances should be asked to present or comment on their findings. The drawback is the inevitable avalanche of correspondence that follows. But if this does happen, perhaps it merely confirms that the reporters or their researchers chose a story with public appeal.

However, having obtained the British Migraine Association's generous offer to answer any correspondence that might follow I accepted an invitation to appear on the United Kingdom television programme *A-plus*. More than two thousand letters were received in the first week from writers of all ages and walks of life. Some passed on useful information about feverfew, but the majority bore sad testimony to the inadequate medical treatment available for migraine – nearly all who wrote wanted to try feverfew.

I was faced with the fact that not only had hopes been raised by the television programme but, as a result of it, thousands more individuals might put themselves on an incompletely tested medicine – something I had always attempted to discour-

'. . . what is taken for today's unorthodoxy is probably going to be tomorrow's convention.' Prince Charles, President of the British Medical Association.

age. It seemed to me that reporters only wanted to discuss the merits of a natural medication such as feverfew on the one hand and the disadvantages of synthetic medicines on the other. I had sometimes found it necessary to prompt interviewers about the very existence of side-effects with feverfew. Very few newspaper and magazine articles mentioned cautionary points, with the exception of *The Times* and the *Sunday Mirror* and *Chemistry in Britain*. In my own article I advised doctors to complete the yellow cards used for the reporting of drug-induced adverse reactions to the CSM whenever they suspected an adverse reaction to feverfew.

Green pharmacy, as herbal remedies are collectively described, is a major growth industry in the UK. Any pharmaceutical distributor will confirm that this is the area of treatment expanding the fastest. Many of these treatments bypass the requirements of the *Medicines Act* by being offered as 'food supplements', and feverfew is no exception.

Many patients now consult alternative (that is, unorthodox) practitioners and a recent survey of general practitioner trainees revealed that seventy out of eighty-six wanted training in techniques such as hypnosis, acupuncture, homoeopathy and herbalism and twelve had actually referred patients for treatment to non-medical practitioners.

Many doctors are open minded about what they, and their patients, see as alternatives to sometimes dangerous drugs but I would tend to the view recently expressed in the *British Medical Journal* by Dr Tony Smith, Deputy Editor, that new treatments should not be used unless validated by clinical trial. This is what I have endeavoured to do at the City of London Migraine Clinic by subjecting feverfew to a comparison with placebo in a double-blind clinical trial.

Feverfew has been used by the majority of migraine sufferers I have met, as a last resort, and usually with some initial scepticism. The success rate seems remarkably high considering that these patients have for the most part failed to respond to orthodox medicines.

Would I give it to a member of my own family? The answer is probably 'yes' if the condition was incapacitating and resistant to other tested medicines. That is, I would consider feverfew *only when all else had failed*. I wouldn't give it to one of my children under twelve or, for reasons given on page 89, to a pregnant member of my family. Nor would I advise feverfew for any person who had developed a rash on contact with the plant as such hypersensitive individuals might develop a much more serious skin disorder if the offending chemicals in feverfew got into the blood stream. Also, instead of being taken as the raw leaf I would prefer to see it given in a capsule or tablet formulation produced by an approved manufacturer in accordance with good pharmaceutical manufacturing practice. This would ensure a constant dosage, which leaves of variable size and chemical content harvested at different times of year, cannot. There is a product, just becoming available, that is a controlled strength feverfew capsule. It differs from other capsules in containing a special preparation of feverfew in oil designed to give a constant dose of feverfew constituents.

I still have qualms on the question of safety. No toxicity tests have been made in animals and so, despite the fact that it has been used for centuries, we still know little about its long-term safety. The blood tests in long-term users do, however, provide grounds for optimism.

Does natural imply safe?

Feverfew users commence taking the plant believing rather than knowing it to be safe, and it comes as a surprise that side-effects sometimes occur. Yet less than half a century ago, many drug treatments used to be obtained from plants and these plants, or chemicals and extracts derived from them, were used to alter disease processes in exactly the same way that many synthetic chemicals do today. Thus, just as the synthetic chemical is defined as a drug when used as a treatment, so is the medicinal plant when used as a treatment. And we have ample evidence in the case of feverfew that this includes the ability to induce side-effects.

It is sometimes claimed that plants and their extracts are less toxic than synthetic compounds simply because they are 'natural', as though their natural occurrence bestows on them a property of safety not enjoyed by synthetic drugs. In fact, plants and plant extracts are simply large collections of chemicals – mini drug firms. These chemicals are just as likely as their synthetic counterparts to cause toxic effects. Indeed, some plant chemicals are highly poisonous: strychnine, hyoscine, muscarine, and curare being well-known examples.

It may be that as more is learned of feverfew, the phenomenon of synergism – where the effect of the whole plant or a crude extract has a stronger or better tolerated effect than one isolated pure constituent – may be demonstrated, but there is no evidence for this at the moment. On the other hand it might be argued that to give an unknown but large number of foreign chemicals is likely to place a greater stress on the body's metabolic processes than would a single chemical whose effect had already been identified in animal studies. It could also be argued, as I have done, that the variable constituents of the leaf from plant to plant and season to season might give an unstable dosing regimen likely to cause problems.

No drug has yet been invented that will not have a damaging effect somewhere in the body if the dose is high enough. Thus we can say that all drugs are toxic when taken in overdose; and in the case of feverfew this toxic dose-level has not been determined. We know from the herbals that up to 140 times the current daily dose was taken for some conditions with little more than a purging effect. But the writers of the herbals might have been reluctant to report their failures.

As Dr R.G. Penn, Principal Medical Officer at the UK Department of Health and Social Security, wrote in the October 1983 number of the *Adverse Drug Reactions Bulletin*:

> The belief that herbal medicines are non-toxic may be badly mistaken, and it is unsafe to rely on the observations and conclusions of patients and physicians of the distant past who would almost certainly have overlooked such subtle toxicities as carcinogenicity, mutagenicity, and hepatotoxicity even though they may have recognized more obvious acute adverse effects. Regrettably, there have been few scientific studies of the efficacy and toxicity of herbal medicines, and most of the relevant reports are of the uncontrolled 'clinical impression' type.

Attitudes of doctors to feverfew

Doctors are constantly being told by their patients of the benefits they have derived from 'alternative' treatments. And a recent study established that one-third of patients with rheumatoid arthritis and two-fifths of those with backache had consulted practitioners of alternative medicine. Indeed, my first clinical encounter with feverfew was through a patient who had sought help from an 'unqualified' person, in this case another feverfew user.

Although the use of medicinal plants was part of orthodox medical practice even in the first half of the twentieth century the use of raw plants by doctors is so uncommon in western medicine today that the same polarization of views among doctors towards the use of feverfew might be expected as towards alternative medicine generally. Either it would be cautiously welcomed by those prepared to await evaluation in a condition where other treatments are demonstrably inadequate or dismissed out of hand as worthless nonsense by those suspicious of any unusual treatment suggested by patients. Indeed knowing how mud can stick to those associated with unusual treatments contributed to my own initial reluctance to commence a feverfew investigation.

In the event, so far, there has been only a positive reaction by doctors. Several general practitioners have volunteered as subjects for further clinical trials, others have offered their services as trial monitors.

Newtownards, Co. Down

As a migraine sufferer and a GP I am interested in feverfew and migraine. I understand very little about the role of the herb (or how to use it) and would value your help and advice. I would be interested in investigating its use in willing subjects if you could send me details of how I could help you in your research.

Incidentally, I have only been able to obtain granulated feverfew locally – to drink – perhaps you could advise me about sources of supply also.

I look forward to hearing from you and helping if possible.

100

Melbourne, Victoria

I am a medical practitioner in Australia and I recently read an article in our local paper here in Melbourne, *The Age*, quoting your recent study on migraine treated with the herb feverfew.

I am very interested indeed in trying this on some of my patients and I would like further information both on the study you have done and on the herb itself, where I can get it and in what doses I should be using it for patients with migraine.

I do hope you will be able to help me.

Some doctors, I discovered, were already taking feverfew – among them a consultant surgeon, consultant psychiatrist and a university professor as well as those in general practice. Some have completed the questionnaire and it appears that they have responded in a similar way to the rest of the population. In the words of a retired consultant surgeon:

Alnwick, Northumberland

I have been taking feverfew daily for a number of years and have reduced the incidence of my attacks from two or three a week to one or two a year. I dry the leaves and then powder them, then take about half a teaspoonful a day.

Some patients are actually being advised by hospital consultants to try feverfew. One doctor who had suggested feverfew had apparently not explained the treatment regimen adequately to his patient, as she wrote:

I have suffered for years from migraine. My doctor has prescribed everything there is on the market for this problem but I can take only one medication – namely Migril. However, at a recent visit to his surgery he suggested I try chrysanthemum leaves. I cannot obtain them here in Aberdeen, only in crystal form and I believe this is not so effective.

Yesterday in our local paper I read the article on migraine and decided to ask your help in obtaining the leaves.

Doctors, including consultant obstetricians, neurologists, psychiatrists, radiologists and surgeons, especially those who were themselves migraine sufferers, were among the first to ask for more information on feverfew, as were hospital administrators, university professors, medical Vice-Presidents of pharmaceutical houses and so on, and these enquiries came from many parts of the world. It is to be hoped that this worldwide interest will generate more clinical studies of ethically acceptable and scientifically sound standards.

And tomorrow?

Much more work remains to be done to convert feverfew from a therapeutic curiosity to an ethical pharmaceutical. In addition to continued clinical vigilance regarding safety aspects, more controlled clinical trials are necessary.

Eventually the long-term safety studies will have to be made in animals. These can only be undertaken by a major pharmaceutical company willing to make a considerable financial commitment.

More information is required on the precise action of the chemicals in feverfew. This is one of the areas in which my colleagues and I expect to maintain our interest – grant-awarding bodies willing!

Meanwhile a number of feverfew capsules, tablets and tisanes continue to be available through the 'health' trade and some contain high doses which have never been properly tested clinically prior to marketing.

So do be careful when purchasing herbal products from a 'health' food outlet or by mail order. Just because they are available doesn't mean they have been tested for efficacy or safety in man or animals. This is nearly always so if a clear medical use for the product does not appear on its label. You can be fairly sure that if any testing has been done the manufacturer will mention it, so read the product information carefully before you buy.

It would be tragic if the promising feverfew project were to founder through an unsuspecting individual suffering an adverse reaction after taking an untested high dose preparation. It is therefore encouraging to learn that a reputable manufacturer has emerged intent on seeing the project progress through to the granting of a full product licence, the current hallmark of quality, efficacy and safety.

Appendix:
The Feverfew Questionnaire

If you would like to assist with the research into Feverfew and have taken the plant for at least one month, fill in the following confidential questionnaire and return it to:

Sheldon Press,
Holy Trinity Church,
Marylebone Road,
London NW1 4DU

(It will be forwarded unopened, to the City of London Migraine Clinic)

and mark the envelope Migraine Questionnaire.

General Information

Surname of user/ex-user ....

Forenames ....

Address...

....

....

Age............................   Sex   M $\boxed{0}$ F $\boxed{1}$

Occupation..

Name of your doctor...

Address...

....

Date..

Do you agree with your doctor being contacted by the City of London Migraine Clinic in confidence for confirmation of diagnosis if necessary?

yes $\boxed{0}$ no $\boxed{1}$

Please tick the appropriate answers in the boxes provided:

1. How did you first learn of feverfew?

From (a) a newspaper/magazine article ☐ 0

(b) a feverfew user ☐ 1

(c) a doctor ☐ 2

(d) a migraine charity ☐ 3

(e) a friend/relative/acquaintance ☐ 5

(f) TV or radio programme ☐ 6

(g) other (specify) .... ☐ 4

2. On the advice or initiative of whom do (did) you take it?

(a) Own ☐ 0

(b) A feverfew user ☐ 1

(c) A Doctor ☐ 2

(d) A friend/relative/acquaintance ☐ 4

(e) Other (specify)... ☐ 3

☐ 5

3. From where do (did) you obtain your supply of feverfew plants?

(a) Own or friend's garden ☐ 0

(b) Nurseryman ☐ 1

(c) Herbalist or health food shop ☐ 2

(d) Countryside ☐ 4

(e) Feverfew user ☐ 5

(f) Waste/neglected land ☐ 6

(g) Other (specify) .... ☐ 3

☐ 7

4 How certain are you that the plant is/was common feverfew?

(a) Absolutely certain ☐ 0

(b) Fairly certain ☐ 1

(c) Uncertain ☐ 2

5 What part(s) of the plant do (did) you eat?

(a) Leaves ☐ 0

(b) Stems ☐ 1

(c) Flowers ☐ 2

(d) Roots ☐ 3

(e) Leaves and stems ☐ 4

(f) Leaves, stems and flowers ☐ 5

(g) Feverfew in tablets/capsules/tea bags ☐ 6

(Please send a specimen of your plant with flower if possible)

6 How do (did) you prefer to eat them?

(a) Fresh ☐ 0

(b) Dried ☐ 1

(c) Doesn't matter ☐ 2

(d) Frozen ☐ 4

(e) As a tea ☐ 5

(f) Other (specify) .... ☐ 3

7. Do (did) you take feverfew

(a) with food? ☐ 0

(b) with water (milk, tea, or coffee)? ☐ 1

(c) without food and drink? ☐ 2

(d) with food and drink? ☐ 3

(e) with or without food and drink? ☐ 4

8. Do (did) you take feverfew every day?

(a) Yes ☐ 0

(b) No ☐ 1

If the answer is no please state how often you take it (e.g. once/twice a week etc):

....

9. How many leaves do (did) you take every day? (Estimate in the equivalent of small leaves possessing five leaflets and measuring approximately 1½ x 1¼ inches).

(a) One | 0

(b) Two | 1

(c) Three | 2

(d) Four | 3

(e) Five or more | 4

(f) Less than one | 5

| 6

10 Do (did) you take this amount

(a) at one time? | 0

(b) spaced out through the day? | 1

11 At what time(s) of day do (did) you take it?

(a) On waking | 0

(b) Breakfast time | 1

(c) Mid-morning | 2

(d) Lunch time | 3

(e) Afternoon | 4

(f) Evening meal time | 5

(g) Evening | 6

(h) On retiring | 7

(i) Other (e.g. more than 1 time)............................... | 8

(j) Anytime | 9

12. What happens (happened) if you forget (forgot) to take a day's supply?

(a) Nothing | 0

(b) If something, specify .... | 1

(c) Never forgot | 2

(d) Not taking feverfew now | 3

13. How long have you taken (did you take) feverfew?

    (a) 1-3 months   `0`

    (b) 4-6 months   `1`

    (c) 6-12 months   `2`

    (d) 1-2 years   `3`

    (e) 2-5 years   `4`

    (f) More than 5 years   `5`

    (g) Can't remember/answer   ☐

14. Have you ever stopped taking feverfew?

Yes `0`   No `1`

When.................................(date)

Why? ....

....

....

15 Does (did) the plant cause you to have any change in the following?

	No_0	Yes_1	If yes please state details how
(a) Urine or bladder control			
(b) Bowel habit			
(c) Appetite			
(d) Sleep			
(e) Mood			
(f) Breathing			
(g) Heart beat			
(h) Weight			
(i) Other (specify)			

16. What unpleasant side-effects do (did) you associate with taking feverfew?

(a) None 0 (b) State: 1	How soon after starting feverfew did each appear?	If you continue taking feverfew do these	
		continue?0	disappear?1
1			
2			
3			
4			
5			

17. What pleasant effects did you notice when you took feverfew?

(a) None 0 (b) State 1	How soon after starting feverfew did each appear?	If you continue taking feverfew do these	
		continue?0	disappear?1
1			
2			
3			
4			
5			
6			

Continue on separate sheet if necessary.

18. **Have you ever noticed or suffered any of the following symptoms or conditions while taking feverfew?**

	No_0	Yes_1	When? Month/Year	Due to feverfew No_0	Due to feverfew Yes_1	Details
(a) Constipation						
(b) Diarrhoea						
(c) Hot flushes						
(d) Irregularity of the pulse (palpitations)						
(e) Mouth ulcers						
(f) Indigestion, heartburn or belching						
(g) Skin rash						
(h) Headache						
(i) Jaundice (yellowing of eyes/skin)						
(j) Loss of appetite						
(k) Bleeding gums						
(l) Swollen ankles						
(m) Excessive thirst						

19 Have you ever taken any other herbal remedies for any medical condition?

(a) No ☐ 0

(b) Yes ☐ 1

If yes, specify ....

....

....

20 When you took feverfew did your migraine headache become

(a) less frequent? ⬚ 0

(b) more frequent? ⬚ 1

(c) less painful? ⬚ 2

(d) more painful? ⬚ 3

(e) unchanged? ⬚ 4

(f) less frequent and less painful? ⬚ 5

(g) nonexistent (i.e. disappear altogether) ⬚ 6

(h) shorter in duration ⬚ 8

⬚ 9

20a Did you suffer from nausea and/or vomiting during your migraine attacks *before* you started taking feverfew?

Nausea

(a) Never ⬚ 0

(b) Rarely ⬚ 1

(c) Occasionally ⬚ 2

(d) Usually ⬚ 3

(e) Always ⬚ 4

Vomiting

(a) Never ⬚ 0

(b) Rarely ⬚ 1

(c) Occasionally ⬚ 2

(d) Usually ⬚ 3

(e) Always, once only ⬚ 4

(f) Always, repeated vomiting ⬚ 5

20b Do/Did you suffer from nausea and/or vomiting during your migraine attacks after you started taking feverfew?

Nausea

(a) Never ⬚ 0

(b) Rarely ⬚ 1

(c) Occasionally ⬚ 2

(d) Usually ⬚ 3

(e) Always ⬚ 4

Vomiting

(a) Never ⬚ 0

(b) Rarely ⬚ 1

(c) Occasionally ⬚ 2

(d) Usually ⬚ 3

(e) Always, once only ⬚ 4

(f) Always, repeated vomiting ⬚ 5

110

20c Since starting feverfew has (did) the nausea associated with your migraine attacks

(a) lessened in frequency? `0`   (d) increased in severity? `3`

(b) increased in frequency? `1`   (e) remained the same? `4`

(c) lessened in severity? `2`

After starting feverfew do (did) you vomit

(a) More often? `0`   (b) Less often? `1`

(c) About the same? `2`

21 If you have now (or ever) stopped taking feverfew did the headaches return? If so, how soon after stopping?

(a) One day `0`   (f) Not applicable (haven't stopped taking feverfew) `5`

(b) Within 1 week `1`

(c) Within 1 month `2`   (g) Headaches never stopped `6`

(d) Within 3 months `3`   (h) Other (specify) `7`

(e) Never `4`

22 What medicines have you ever taken for your migraine?

(a) None `0`   (b) See below `1`

Please write name of one medicine by each letter:	Does/did it help?		Do you take it now?		If not, why not?
	No$_0$	Yes$_1$	No$_0$	Yes$_1$	
(a)					
(b)					
(c)					
(d)					
(e)					
(f)					
(g)					

23 Are you suffering from any other illness or condition requiring medical treatment or regular visits to your doctor at the present time?

(a) No ☐ 0

(b) Yes ☐ 1

If yes please give details including the medicines you take

....

....

....

....

24 How old were you when your migraine began?.............years.

25 How many attacks of migraine did you suffer on average before you started treatment with feverfew?

(a) Less than 1/month ☐ 0

(b) 1-2/month ☐

(c) 3-6/month ☐

(d) 2/week ☐

(e) 3-4/week ☐

(f) 1/day or night ☐

☐ 9

26 How many attacks of migraine do (did) you get while taking feverfew?

(a) None ☐

(b) Less than 1/month ☐

(c) 1-2/month ☐

(d) 2/week ☐

(e) 3-4/week ☐

(f) 1/day or night ☐

☐ 9

112

27 Who diagnosed your headaches as migraine?

    (a) You   `0`

    (b) Your doctor   `1`

    (c) Other person (specify) .... `2`

28 Which of the following diagnoses has/have been applied to your migraine?

    (a) Common migraine   `0`

    (b) Classical migraine   `1`

    (c) Tension migraine   `2`

    (d) Cluster headache   `3`

    (e) Common and classical   `7`

    (f) Abdominal migraine   `9`

    (g) None of these   `4`

    (h) Don't know   `5`

    (i) Other (specify) .... `6`

                        `8`

29 Does anything in particular bring on a migraine attack?

No `0`   Yes `1`

    If yes, please give details ....

....

....

....

Please complete this section only if you have *ever stopped* taking feverfew.

30 For how many months did you take feverfew?

....

31 During the two months after you stopped taking feverfew did you experience any of the following?

	No$_0$	Yes$_1$	When (weeks after stopping feverfew)	Please give details
(a) Mouth ulcers				
(b) Headache or migraine attacks				
(c) Pain in the legs				
(d) Stiffness of muscles/joints				
(e) Painful muscles/joints				
(f) Pain or cramp elsewhere (specify)				
(g) Influenza-like illness or rheumatism				
(h) Change in bowel habit (specify)				
(i) Excessive thirst				
(j) Increased urine production or getting up at night to pass urine				
(k) Indigestion, heartburn or belching				
(l) Skin rash				
(m) Change in your periods (if applicable)				
(n) Change in weight				
(o) Change in appetite				
(p) Weakness of hand grip				

(q) Loss of sensation in arm or leg .				
(r) Pain 'flitting' from place to place				
(s) Other (specify)				

32 Did you have to take any medicines during the two month period you stopped taking feverfew?

        (a) No   &boxed;0

        · (b) Yes   1   (state) ....

        (c) Reason ....

        (d) How long after stopping? ....

33 How soon after stopping feverfew did your headaches return?

............days; ....weeks; ....months; .... did not return.

34 During the first months after stopping feverfew how did your headaches compare with those occurring *before* feverfew:

      (a) more frequent?   `0`

      (b) less frequent?   `1`

      (c) less painful?   `2`

      (d) more painful?   `3`

      (e) unchanged?   `4`

      (f) headache didn't return?   `5`

      (g) more frequent and more painful?   `6`

      (h) more frequent and less painful?   `7`

      (i) less frequent and more painful?   `8`

      (j) less frequent and less painful?   `9`

                                             `10`

35 During the second month after stopping feverfew how did your headaches compare with those occurring *before* feverfew:

(a) more frequent? `0`

(b) less frequent? `1`

(c) less painful? `2`

(d) more painful? `3`

(e) unchanged? `4`

(f) headache didn't return? `5`

(g) more frequent and more painful? `6`

(h) more frequent and less painful? `7`

(i) less frequent and more painful? `8`

(j) less frequent and less painful? `9`

`10`

36 During the third month after stopping feverfew how did your headaches compare with those occurring *before* feverfew:

(a) more frequent? `0`

(b) less frequent? `1`

(c) less painful? `2`

(d) more painful? `3`

(e) unchanged? `4`

(f) headache didn't return? `5`

(g) more frequent and more painful? `6`

(h) more frequent and less painful? `7`

(i) less frequent and more painful? `8`

(j) less frequent and less painful? `9`

`10`

116

Index

117